OSMOTIC INVESTIGATIONS

OSMOTISCHE UNTERSUCHUNGEN.

STUDIEN ZUR ZELLMECHANIK

VON

DR. W. PFEFFER,
PROFESSOR DER BOTANIK IN BASEL.

MIT FÜNF HOLZSCHNITTEN.

LEIPZIG,
VERLAG VON WILHELM ENGELMANN.
1877.

OSMOTIC INVESTIGATIONS

STUDIES ON CELL MECHANICS

by

Dr. W. PFEFFER
Professor of Botany at Basel

with five figures

Leipzig, Wilhelm Engelmann, 1877

translated, with an introduction, by

G. R. Kepner and Ed. J. Stadelmann

University of Minnesota

VNR VAN NOSTRAND REINHOLD COMPANY
New York

Library of Congress Catalog Card Number: 84-25774
ISBN: 0-442-23007-9

Manufactured in the United States of America.

Published by Van Nostrand Reinhold Company Inc.
135 West 50th Street
New York, New York 10020

Van Nostrand Reinhold Company Limited
Molly Millars Lane
Wokingham, Berkshire RG11 2PY, England

Van Nostrand Reinhold
480 Latrobe Street
Melbourne, Victoria 3000, Australia

Macmillan of Canada
Division of Gage Publishing Limited
164 Commander Boulevard
Agincourt, Ontario MIS 3C7, Canada

15 14 13 12 11 10 9 8 7 6 5 4 3 2 1

Library of Congress Cataloging in Publication Data
Pfeffer, W. (Wilhelm), 1845-1920.
Osmotic investigations.
Translation of: Osmotische Untersuchungen.
Includes bibliographies and index.
1. Osmosis. I. Title.
QH615.P4413 1985 574.87'5 84-25774
ISBN 0-442-27583-8

FOREWORD

In an endeavor to relate certain processes of motion to the cell mechanism underlying them, I discovered facts whose causes first had to be explained before I could expect to make any further progress. Above all, it was necessary to discover the cause of the conspicuously high hydrostatic pressure that exists even in plant cells whose cell sap is only a dilute solution. Here, the way I formulated the questions for experimental research was derived from observations on the plant cell. My next question was, what osmotic pressure is produced by solutes and, specifically, the so-called crystalloids, when they do not diosmose? Traube's precipitation membranes, using the plant cell as a guide, made it possible for me to construct the apparatus that I used in the studies described in the physical part.

Having only physiological considerations in mind, I occasionally stopped short at the very points where a physicist, from his point of view, would have most wanted to direct his focus of attention. But my task as a physiologist was to look for the fundamental principles relevant to the cell mechanics [*Zellmechanik, i.e. mechanism as seen by the physicist in terms of mechanics*] of osmotic processes with the help of experiences gained in physics. Basing my work on these principles, I then examined special physiological phenomena in order to put the present state of affairs in clearer terms. I hope this will provide impetus for others to become active in an area in which a single researcher's energies cannot hope to be adequate; for osmotic processes are involved in almost all question related to metabolism and energy transduction within the organism.

Bonn, November 1876. W. Pfeffer

CONTENTS

I. Physical Part

A. Apparatus and Methods

B. Experiments and Conclusions

II. Physiological Part

INTRODUCTION

We believe that scientists who study membrane structure and function will find Pfeffer's ideas on these subjects intriguing, because of their unusual scope and their congruity with modern views. These ideas evolved during the years he carried out his pioneering work on osmotic pressure measurements and appeared in his classic monograph, Osmotische Untersuchungen (1877). Some three decades later, Overton (1907:765) would label this work "epoch-making." Yet, as we shall show here, Pfeffer has still received less than his due. Thus, this translation -- which uses Overton's personal copy of Pfeffer's book, replete with marginalia in his own hand, as text source -- has been undertaken not only to disseminate Pfeffer's major contributions, but also to make available his authoritative and historically interesting commentaries on the mid-19th century struggles that advanced our understanding of the cell membrane and osmotic phenomena.

Part one of this introduction summarizes some of the key ideas presented in Pfeffer's book, including those lesser known, and offers comments on why these ideas apparently went unrecognized at the time and later. In part two, after briefly sketching the important events in Pfeffer's early scientific development, we expand this summary into a more detailed description of Pfeffer's ideas regarding the cell membrane and osmosis, as seen in the context provided by the scientific climate of his time. It also includes occasional remarks relating Pfeffer's ideas to subsequent developments in the 19th and 20th centuries.

1. In the case of the plasma membrane, Pfeffer effectively argues that at each surface of the plant cell's protoplasm there must exist a limiting membrane, whose major constituents are protein. Much less well known is his idea that membranes are formed from structural units, called tagmas, which, for biological membranes, consist of protein or colloid-like protein aggregates. Accordingly, the tagmas provide a structural framework that enables Pfeffer to sort out the paths that solute and water molecules take through membranes: first, the intertagmatic pathways include water-filled pores and diffusion paths; second, the amphitagmatic pathway allows for diffusion at adjoining tagma surfaces; third, the diatagmatic pathway involves movements through the tagmas themselves. Moreover, since Pfeffer expects pores to exhibit different sizes, shapes and lengths, membranes must behave like sieves with respect to solute permeability. In addition, such concepts as solvent drag, the unstirred layer and one-way permeability gates appear in this book.

Pfeffer's accurate measurements of osmotic pressure eventually received considerable recognition after van't Hoff used them to support his theory of osmotic pressure. Still, there is much more. For instance, in the case of permeable solutes, Pfeffer defines the osmotic pressure as the equilibrium pressure that balances osmotic inflow with filtration outflow and relates this pressure to the relative sizes of the membrane pore and the permeable solute molecule, thus foreshadowing the notion of the reflection coefficient, σ.

Equally instructive are the interesting mechanisms he develops to explain osmotic pressure and osmosis at the level of molecular interactions. One, rooted in the newly emerging kinetic theory, pictures osmosis as resulting from the different number of water molecule impacts per unit area on the membrane, when unequal concentrations of water molecules reside on each side. The other expounds a mechanism based on the different attractive forces among the membrane tagmas, the solute molecules and the water. Thus, at the membrane:solution interface, a so-called solute diffusion zone develops, whose concentration gradient influences water flow into and out of the membrane. Furthermore, the idea that the number of solute molecules per unit volume provides the driving force for osmosis comes from Pfeffer, who also analyzes why this relationship departs from ideality for non-dilute solutions. However, in general, even the renewed interest in Pfeffer's work occasioned by van't Hoff's theory did not lead to greater appreciation of the ideas described above.

Further consideration of why so much that Pfeffer offered

in Osmotische Untersuchungen was missed, or dismissed, and why the book was never translated reveals several factors which, taken together, help to explain this neglect. First, Pfeffer's ideas are often speculative and abstract, especially those derived from kinetic theory and mechanics; thus, the botanists and biologists who ought to have been interested in his work would certainly have found it hard going. Second, these difficulties are accentuated by Pfeffer's abstruse and ambiguous writing style, which no doubt partly reflects the struggle to clarify his own thinking. Third, the physicists and chemists had, by 1877, largely abandoned their earlier intense interest in osmosis, which contributed to the lack of a receptive audience for these ideas. Moreover, Pfeffer's interest in membranes proved to be untimely because the then widely accepted view was that of Max Schultze (1861), who asserted that animal cells, and by implication plant cells, did not possess a plasma membrane. With the passage of decades, these factors were further compounded by the newly developing perspectives that led scientists away from the ideas in the older literature, even though one occasionally finds in the early 20th century a reference to the broad scope and originality of Pfeffer's work (Collander, 1924) -- still, almost all of it became lost to sight.

2. Wilhelm Pfeffer (born March 9, 1845 in Grebenstein, Germany) was to be strongly influenced towards botanical science by his boyhood experiences with herbs and medicinals in his father's apothecary (Bünning, 1975). Although Pfeffer actually studied chemistry at Göttingen, taking his Ph.D. in 1865, his subsequent post-doctoral years were spent in the laboratories of various leading botanists, including Pringsheim and Sachs. In fact, they introduced him to the research problems in plant physiology that eventually provided the practical motivation for his undertaking the study of osmotic and membrane phenomena.

In early 1871 Pfeffer completed his Habilitation thesis at Marburg and thus became Privatdozent. During 1871-72, while still in Marburg, he published several papers and pursued the study of stimulation movements in plants. The problems therein, along with his growing curiosity about what caused the high hydrostatic pressures observed in plant cells, led him to conduct preliminary experiments on osmotic pressure in 1872. Simultaneously, he continued to study the "mechanics" of the cell, to seek the causes of metabolism and energy transduction, and to move closer to applying

Helmholtz's concept of a general physiology to the study of cells. Indeed, Pfeffer considered Helmholtz to be the model scientist and dedicated his first book, Physiologische Untersuchungen (1873), to him. That same year also saw Pfeffer take up his first professorial appointment, at Bonn, where he would pursue in earnest the study of osmotic phenomena.

Pfeffer brought to that task three especially useful attributes. His practical interests strongly motivated him to venture this excursion into fundamental research because he considered an understanding of osmosis to be essential to proceeding further with the physiological problems he was pursuing. His ambition, combined with the enthusiasm of a newly-minted young professor, allowed him to commit his considerable energies to what would turn out to be a project of several years duration. Finally, his training in chemistry would prove to be excellent preparation for someone attacking a problem that required these skills: constructing a unique apparatus for measuring osmotic pressure across membranes, mastering the chemical reactions used to form artificial membranes, and delving into molecular interaction theory to explain osmotic mechanisms and membrane structure.

Fortune also played a role, in the form of propitious timing. The resources of German society had earlier been diverted toward the struggle for German unification, which culminated with Bismarck forging the German state in 1871. Thereafter, the enterprises of university researchers benefited from a greater availability of resources (McClelland, 1980).

Eventually, the results of Pfeffer's labors, carried out in Bonn from 1873-76, were published in Osmotische Untersuchungen, in 1877, the year in which Pfeffer left Bonn to become a full professor at Basle. In the book's foreword, the now older and wiser professor expresses the view that the subject of osmotic phenomena is too vast for a single researcher's energies and hopes that the book will provide an impetus for others to undertake such studies. He surely could not have known at the time that this would also be his major contribution to the field. His life's work in plant physiology awaited and would take him away from experimental work on osmotic pressure, though he would continue to comment on the subject in subsequent textbooks and review articles.

Before taking up Pfeffer's ideas on membranes and osmosis in more detail (this discussion omits other interesting topics in Pfeffer's book, e.g. tropisms and stimulation movements), we would like to sketch the climate of mid-19th century thought regarding osmotic phenomena. By then, osmosis was at least a century-old observation that had been repeated numerous times using a variety of organic (e.g. animal bladders, epithelia) and inorganic (e.g., porcelain plates, egg shells) materials [see Partington (1964) for the earlier references]. Pfeffer reviews the older work, identifies the major issues and offers intriguing glimpses into the character of certain key researchers, particularly Dutrochet.

The basic apparatus employed in most earlier studies was the simple endosmometer of Dutrochet. Usually, this took the form of an inverted funnel with the mouth covered by a diaphragm, or "membrane," of suitable material. It was then placed in a water bath (outside) and the funnel filled with a solution (inside) containing the test solute. Observation revealed a fluid stream flowing through the membrane from outside to inside, the endosmotic flow. Invariably, it was accompanied by a so-called stream of solution moving from inside to outside, the exosmotic flow, which resulted from the fact that the materials used as membranes were permeable to the test solutes commonly used. The changing fluid level in the stem of the inverted funnel served to measure the osmotic effect, or osmotic pressure, of the inside solution.

In such a system, using an animal bladder to separate water on the outside from an alcohol solution inside, experimenters observed that the water current inward (endosmosis) exceeded the alcohol current outward (exosmosis), so the fluid level inside rose. Whereas, using a collodion membrane, the alcohol current exceeded the water current and the fluid level inside fell. These and other observations led various investigators (e.g., Dutrochet and Graham) to suggest that the "affinity" of a solution for a particular membrane determined the relative magnitudes of the solution current and water current through that membrane. Thus, in the case of a saline solution separated from water by a membrane, an exosmotic current moved towards the water and an endosmotic current moved towards the saline. Eventually, Graham (1854:178) resolved this double current question by recognizing that the salt molecules diffused independently of water. Consequently, he labelled the pure water current, moving towards the saline solution, osmosis.

The variable behavior towards different solutes of the organic membranes, whose properties often changed over the long time course of the typical experiment, and the absence of reliable quantitative data led to considerable disputes over the various theories of osmosis that were put forth.

Often, these theories assumed that capillary-like structures, pores, existed in membranes. One particularly significant approach originated with Brücke in the early 1840's and later was developed more fully by Fick. These investigators assumed that the differing solute and water interactions with the pore walls provided the basic mechanism regulating the flows through the pore. Equally important was the concept, put forth by the physicist Philipp Jolly, that the amount of water moving inward divided by the amount of solute moving outward was a constant, for a given solute and membrane, called the endosmotic equivalent of the solute. Although Jolly provided data supporting his theory, other workers obtained contradictory results suggesting that the nature and condition of the membrane influenced the results. Interestingly, Pfeffer discusses these approaches in the informative historical review of osmosis that he presents in his book, and explicitly excludes studies of the endosmotic equivalent from his research program. Moreover, sometime later, Ostwald (1891:154) was to conclude that, by the late 1860's, physicists and chemists had taken the view that if osmotic phenomena depended on the nature of the membrane, then they belonged to the realm of the physiologist. In fact, from that time up to Pfeffer, research on osmosis diminished and no significant new theories emerged to explain osmotic phenomena.

Meanwhile, the technical innovation that would eventually enable Pfeffer to advance the understanding of osmotic phenomena was being developed by Traube (1867), namely, artificial precipitation membanes, e.g., of copper ferrocyanide. Traube, however, could not measure the pressures developed across these membranes because he formed them as coatings on drops of solution and the slightest excess pressure overcame their resistance to breakage. The essential feature of these membranes was their virtual impermeability to sucrose, gum arabic, gelatin and certain salts used as test solutes in osmotic studies. Their significance, as Traube indicated, was that it now became possible, in principle, to measure osmotic pressure in the absence of solute flow. Pfeffer appreciated the value of this technique and, more importantly, eventually mastered the formidable difficulties involved in preparing these membranes in the wall of a porous porcelain pot, which provided the mechanical support needed to resist pressures reaching several atmospheres. Thus, he acquired the capability for studying osmotic phenomena in a quantitative way.

What would become the best known results of Pfeffer's

work, the accurate measurement of the osmotic pressure that an impermeable solute produced across the copper ferrocyanide membrane and the finding that osmotic pressure is proportional to concentration and to temperature, had to await the appearance of van't Hoff's theory before their significance could be appreciated. That theory relied on two key developments in chemistry, one of which ought to have been known to Pfeffer and the other of which could not. The van der Waals equation for gases appeared in 1873, but Raoult's law of partial pressures was only developed in the early 1880's. The latter, especially, led to the recognition of the need to employ mole quantities in describing certain properties of solutions. Together with Pfeffer's osmotic pressure measurements, these insights contributed to van't Hoff's formulating the osmotic pressure equation. In fact, several decades would pass before Pfeffer's data were superseded by more precise measurements, using improved versions of Pfeffer's technique [a summary of this important work appears in Findlay (1919:6, 12-39)]. Still, the primacy of Pfeffer's work continued to afford him recognition.

What remained obscure, however, was Pfeffer's extensive explication of his own theories for the osmotic pressure mechanism. As becomes apparent in reading Pfeffer's review on osmosis, he actively took up the earlier ideas of Brücke and Fick on pore diffusion. To these, Pfeffer added insights based on his understanding of mechanics and kinetic theory, which led him to develop molecular mechanisms for describing osmosis. For example, his kinetic theory mechanism assumed that osmotic water flow is proportional to the number of water molecules striking a unit area of the membrane in unit time. Thus, on the side of the membrane that sees only pure water, this effect is maximal. Whereas, on the other side, which sees a solution containing solute and water, the solute molecules compete with the water and therefore reduce the number of water molecule impacts. As a result, net water flow occurs toward the solution side.

This picture precedes and differs significantly from the ideas van't Hoff put forth regarding the mechanism of osmotic pressure, which emphasized solute bombardment of the membrane [see van't Hoff's 1887 paper, translated in Kepner (1979:21), and Partington (1964:654)]. Interestingly, according to Cohen (1915:120), van't Hoff treated Pfeffer's book as a personal vade-mecum, though he apparently never cited Pfeffer's kinetic theory mechanism in his own discussions of this topic. However, it should be noted that van't Hoff ascribed little significance to the quest for a mechanism that might explain osmosis, preferring to rely on the thermodynamic description as sufficient for all practical purposes. On the whole, Pfeffer's kinetic picture squares more closely with the thermodynamic approach that developed in the 1890's [see Washburn (1910)], and focused attention on the mole fraction of water, instead of solute concentration, as the source of the osmotic driving force [Mauro (1981) offers a current perspective on the continuing search for a mechanism of osmosis].

Pfeffer then develops a second mechanism, which he considers to be the more important, that almost certainly was necessitated by his experimental data showing that, for equal weight percent solutions, colloids gave higher osmotic pressure than crystalloids, in certain membranes. In fact, the actual source of Pfeffer's difficulties here resided in his implicit assumption that solutions of equal weight percent should have equal osmotic effects, for impermeable solutes. Such an assumption was completely natural for the time, since the importance of mole quantities had not yet been recognized. It should be noted that, subsequently, in the second edition of his textbook on plant physiology, Pfeffer (1900:138, 144 - footnote one) discussed his earlier views and pointed out this misconception.

All this led Pfeffer to elaborate a theory in which solute molecules were attracted to the membrane surface by tagmas. Therefore, a solute diffusion zone formed that extended from the membrane out into the solution. The driving force for water flow now becomes the attraction between solute molecules in this boundary layer and water molecules in the membrane. Thus, the solute concentration at the interface "pulls" water out of the membrane and creates a net osmotic flow across the membrane. Moreover, this mechanism allowed Pfeffer to explain his data on crystalloids and colloids because the profiles of their diffusion zones would differ, due to the differing molecular attractions between these molecules and the tagmas. As a result, he could, in theory, account for the different osmotic pressures observed with such solutions of relatively impermeable solutes. Hence, the essence of this mechanism lies in the assumption that the membrane itself affects osmotic equilibrium because the tagmas interact differently with different solutes; Pfeffer (1900:144) eventually recognized that this too was incorrect.

In the course of describing these mechanisms, he also presents other important ideas about osmotic phenomena. For example, Pfeffer explicitly defines osmotic pressure as the equilibrium state between endosmosis and filtration. He also points out that the osmotic water flow is proportional to the driving force and, by analogy with Fick's law for solute diffusion, he equates the constant of proportionality, k, with the membrane diffusion constant for water. Furthermore, Pfeffer goes on to identify the solute concentration gradient across the membrane as the driving force. In Pfeffer's view, the appropriate measure of solute concentration here is the number of solute molecules per unit volume of solution. On the other hand, he suggests that the postulated linear relationship between flow and solute concentration difference should only hold for dilute solutions. This is so because only then are the solute molecules, in his terms, fully "saturated" by the surrounding water.

At this time, there was no standardized measure of solute concentration among researchers interested in osmosis, as Pfeffer notes in his brief survey of the literature. Thus, Dutrochet had used specific gravity differences, Fick and Beilstein used weight percent, Jolly used grams of solute divided by grams of water, and Voit used volume percent. Later, van't Hoff chose to use volume percent, as did Ostwald and Nernst. Further, all the above-mentioned workers had chosen for their reference volume the solution volume; in fact, the best choice for osmotic pressure measurements would turn out to be the solvent volume. Interestingly, Pfeffer (1900:144, 147) first pointed out that the correct choice is the solvent volume, and even calculated osmotic pressures on this basis, though an extensive discussion of this question in the paper by Morse, Frazer, and Dunbar (1907) does not refer to Pfeffer.

For a permeable solute, Pfeffer analyzes how the equilibrium osmotic pressure developed depends on the relative size of the solute and the pore. Thus, the maximal osmotic pressure developed will decrease when the solute encounters membranes with wider pores. This is, of course, the concept of σ. Pfeffer further notes that the existence of pores means that the pressure head created by endosmosis will, in turn, drive the inside solution, containing water and solute, through the membrane and thus increase the effective solute outflow, i.e., solvent drag [see also Pfeffer (1876:121)]. The narrower the pores, the smaller will be this effect. He adds that if this solute flow is not immediately able to leave the region adjoining the precipitation membrane's exterior surface, located in the wall of the porcelain cell, then it will act to reduce the effective pressure across the membrane; we now call this an unstirred layer effect.

Unfortunately, the breadth and insights contained in Pfeffer's approach to the osmotic mechanism never seemed to stimulate 19th century investigators to pursue this question. Indeed, the line of inquiry that focused on the solute and water interactions within pores, extending from Brücke to Fick to Pfeffer, came to a halt. In the mid-20th century, it would be revitalized and applied, productively, to solute and water flow across biological membranes.

Pfeffer's osmotic studies with artificial membranes also led him to undertake a detailed investigation of the cell plasma membrane. Pfeffer begins part II of his book by analyzing, from his particular viewpoint, the historical development of the then current issues concerning plant cell membranes. This analysis avoids the question of the plasma membrane's existence in animal cells. Of course, such an omission seems prudent in view of the position taken by the animal biologists, described previously, which would dominate cell biology to the end of the 19th century (Kepner, 1983).

What we now call the cell membrane concept experienced a lengthy formative period, extending from Schwann through Pfeffer and on to Overton. In the case of plant cells, the idea arose early on that there ought to be a membrane barrier to regulate exchanges between the cell and the environment. However, identifying the corresponding structural entity proved difficult and so the list of suggestions fluctuated, as did the terminology [see Griffith and Henfrey (1860: 121-128, 579-581), deVries (1877:22-31)]. Thus, among plant physiologists, this concept continued to be subjected to critical discussion. The major contributors to the dialogue included Mohl, Nägeli, Pringsheim and Pfeffer, and their main concerns seemed to be with evaluating the interpretations of the available evidence. For example, the evidence from microscopic studies supported the notion that a barely discernable hyaline layer, of varying thickness, resided along the border of the protoplasm body, the structure that encloses the plant cell sap. Commonly, this layer was identified by the term *Hautschicht* (literally, a skin layer), or some term derived from it. The key feature here, an optically visible layer, stands in contrast to the modern conception, which defines the plasma membrane as sub-microscopic. As a matter of fact, Pfeffer (1873:135-footnote one) had suggested already that the plasma membrane might turn out to be only a molecule thick.

From a functional point of view, observations on the

shrinking and swelling behavior of the protoplasm body and cell sap persuaded some investigators that there must be a membrane barrier at the protoplasm surface. Others, unconvinced, maintained that a not-directly-observable membrane is no membrane (Bünning, 1975:42). Moreover, it was also suggested that the protoplasm body itself functioned as a membrane. Accordingly, the subject fostered ambiguity in terminology and interpretation, that Pfeffer would clarify in his book, where he addressed the crucial issue of the membrane's existence.

This dominates part II, where he presents many experimental observations supporting his main points. The first is the previously mentioned observations of shrinking and swelling seen in the protoplasts of intact cells and in the vacuoles that form spontaneously, in the surrounding solution, after crushing the plant cell's wall and releasing its protoplasmic contents. The second, described in detail, is his studies on dye permeability and impermeability in such vacuoles and intact cells. The last is the need for another barrier, in addition to the plant cell wall, to account for the fact that the hydrostatic pressure in the living plant cell is high, whereas in a dead cell this pressure drops to zero. In fact, his analysis formed the best case, then and for some decades to come, that a plasma membrane existed at any surface where plant cell protoplasm contacted a solution. Indeed, as Pfeffer points out, there has to be a plasma membrane around the cell sap, as well as the protoplasm body. He then goes on to coin the box-within-a-box image that so vividly describes the topological relationship of the sap and the protoplasm in the plant cell.

Although in part I he discusses the pathways through an artificial precipitation membrane composed of tagmas, he clearly extends these ideas to the plasma membrane discussion in part II. Pfeffer does this, in part, by equating the tagma with the protein and protein aggregates that he believes make up the plasma membrane structural elements. Thus, for Pfeffer, the concept of the tagma leads naturally to a membrane structure that offers various pathways for molecular movements through it. For example, his tagmas form water-filled pores between them that allow solute and water to cross the membrane. On the other hand, a solute molecule may enter into combination with the tagma itself to effect its passage through the membrane. Solute and water also find available, at the interfaces of the tagmas, an additional pathway for permeation. For all that, Pfeffer's comprehensive perspective was fated to lapse into obscurity, only to be rediscovered in bits and pieces over the next century and reapplied to biological membranes. On the other hand, one notes that there was no mention of any concept akin to active transport in Pfeffer's book, nor did he consider the possibility that lipids might be a structural element of membranes. This would later involve him in a controversy with the physicist Quincke, who vigorously advocated the necessity for lipids in membranes (Kepner, 1983). On the whole, when considered in the context of the limited information available to him, Pfeffer's description of membrane structure seems remarkably perceptive. Hence, one wonders to what extent our understanding of cell membrane structure and function would have been enhanced by the adoption, early on, of Pfeffer's encompassing perspective as the conceptual paradigm into which subsequent research results would have been fitted. In fact, it seems hard to imagine otherwise, since the actual advance in our understanding was rather modest until around 1925 (Kepner, 1983).

For Pfeffer, the study of osmosis relates intimately to the study of membrane structure. This theme appears throughout the book, especially when he analyzes these phenomena at the level of molecular interactions. The reader witnesses, first-hand, the struggles Pfeffer underwent while sorting out the complexities of these problems. Obviously, the going was often difficult and Pfeffer's views are on occasion fatally flawed; yet, it is fascinating to observe the way he built his arguments and the great care he took to cover virtually every aspect of each issue he considered. Indeed, as Fitting (1920:46) observed, Pfeffer always wanted to secure his position on all sides and was greatly concerned about leaving himself exposed on a weak point. The following comment seems to reveal, in part, why he found himself driven to undertake such labors (p. 33): "Though our understanding of the experimental results does not necessarily compel us to presuppose a very definite conception of the molecular structure of precipitation membranes, we feel even more an intellectual need for deeper insight because only on the basis of such insight can our thoughts follow the path of a solute through the membrane."

Why was it then that Pfeffer seemed to move away from the experimental study of osmosis (see also Czapek, 1921), not even taking it up after van't Hoff's theory emerged to clarify the theoretical and quantitative aspects of the phenomenon? We cannot offer a definite answer, but we do know that his career advanced swiftly after 1877. Within four years he had already become a full professor, head of his own institute in Tübingen, author of the major textbook on plant physiology and an acknowledged leader in the field. Perhaps there was no longer time or energy enough to renew his assault on the many questions raised by his osmotic studies. He was to leave Tübingen in 1887 for Leipzig, where he would continue to study plant physiology until his death in 1920. The next year saw the reissue of his Osmotische Untersuchungen, unaltered -- a fitting tribute to the epochal quality of his contributions, recognized as well as unrecognized.

References

Bünning, E. (1975) Wilhelm Pfeffer, Wiss. Verlagsges., Stuttgart.

Cohen, E. (1915) Die Naturwissenschaften, 3, 118-120.

Collander, R. (1924) Kolloid Chemische Beihefte, 19, 72-105.

Czapek, F. (1921) in the supplementary foreword, p. IX-XIV, to the reissue of Osmotische Untersuchungen, Engelmann, Leipzig.

Findlay, A. (1919) Osmotic Pressure, Longmans, London.

Fitting, H. (1920) Ber. d. Deutsch. Bot. Gesellschaft, 38, (1)-(63).

Graham, T. (1854) Phil. Trans. Roy. Soc., 144, 177-228.

Griffith, J. and A. Henfrey (1860) The Micrographic Dictionary, 2nd ed., van Voorst, London.

Kepner, G. (1979) Cell Membrane Permeability and Transport, Dowden, Hutchinson and Ross, Stroudsburg, Pennsylvania.

Kepner, G. (1983) in Liposome Letters, ed. A. D. Bangham, p. 15-27, Academic Press, London.

Mauro, A. (1981) in Water Transport Across Epithelia, ed. H. Ussing, N. Bindslev, N. Lassen and O. Sten-Knudsen, p. 107-119, Munksgaard, Copenhagen.

McClelland, C. (1980) State, Society, and University in Germany 1700-1914, Cambridge Univ. Press, Cambridge.

Morse, H., J. Frazer and P. Dunbar (1907) Am. Chem. J., 38, 175-226.

Ostwald, W. (1891) Solutions, transl. M.M.P. Muir, Longmans, London.

Overton, E. (1907) in Handbuch d. Physiologie d. Menschen, vol. II, ed. W. Nagel, Vieweg, Brunswick.

Partington, J. (1964) A History of Chemistry, vol. 4, 650-656, Macmillan, London.

Pfeffer, W. (1873) Physiologische Untersuchungen, Engelmann, Leipzig.

Pfeffer, W. (1876) Landwirtsch. Jahrb., 5, 87-130.

Pfeffer, W. (1900) The Physiology of Plants, 2nd ed., transl. A. Ewart, Clarendon Press, Oxford.

Schultze, M. (1861) Archiv f. Anat. u. Physiol. u. wiss. Med., 1-27.

deVries, H. (1877) Untersuchungen über d. mechanischen Ursachen d. Zellstreckung, Engelmann, Leipzig.

Washburn, E. (1910) J. Amer. Chem. Soc., 32, 653-670.

Translation Notes

We have tried to convey clearly Pfeffer's work and ideas, while also retaining much that is characteristic of his style of expression. On occasion, we inserted clarifications of older terms and corrected obvious typographical errors in the original text. Since Pfeffer's description and analysis of membrane structure and function is the most authoritative in the 19th century, and since this terminology had undergone considerable fluctuation up to Pfeffer's time, we chose to insert, at first occurrence, the German term used by Pfeffer to identify each aspect of the membrane concept. Also, format constraints forced us to compile the numerous footnotes and references into sections, instead of including each in toto on the page where it appears. In addition, each reference was checked, its form was modernized, and false citations were corrected where possible with exceptions being denoted by an asterisk in the reference list. Finally, since there was no index in Pfeffer's book, we prepared one for this translation.

We wish to acknowledge the financial support provided for this project by the Department of Physiology, University of Minnesota Medical School. E.J.S. also acknowledges support from Project 80, Agricultural Experiment Station, University of Minnesota. Additionally, we appreciate the valued assistance of Mr. William Kroll, of Krollkraft Inc., who provided a working draft, which we used in preparing the translation. We also made use of the brief excerpts translated by H. Jones (see Harper's Scientific Memoirs, ed. J. Ames, IV, 1899) and M.M.P. Muir (see Ostwald, 1891, in the references for our introduction). The responsibility for the accuracy of the translation presented here is, of course, ours.

I.

Physical Part.

A. Apparatus and Methods.

1. Preparation of the Cells.

Certain precipitates can be obtained as membranes [*Membranen*] if they are formed at the plane of contact of two solutions, or of a solution and a solid. Traube,[1] as is known, was the first to prepare such membranes, and he, at the same time, worked out the conditions under which they can be formed, conditions which later will be briefly explained. The author of this important discovery tested membranes obtained from different substances as to their permeability to dissolved bodies; and it was shown that substances in general pass less easily through such membranes than through those formerly employed in diosmotic investigations. Indeed, many substances which easily diosmose through the latter were incapable of passing through firm precipitation membranes [*Niederschlagsmembranen*].

Traube carried out diosmotic investigations, for the most part, with membranes that closed one end of a glass tube, into which the substance whose diosmotic properties were to be tested was introduced. Such an apparatus is, in most cases, easily prepared; a small quantity of one of the solutions necessary to produce a precipitate being introduced into the glass tube, which is then dipped into the other solution. With correct procedure, the precipitate is formed as a membrane closing the tube, at the surface where the solutions of the two "membrane-formers" come in contact.

Traube worked in every case with cells protruding free into the liquid. These are, on the one hand, not very resistant; and, further, they continually increase in size as long as an osmotic current of water flowing in produces a pressure in the interior which tends to distend the membrane [*Haut*]. By this means a new membrane particle is inserted as soon as the two membrane-formers meet in the enlarged interstices--an increase in surface by intussusception that these membranes very beautifully demonstrate. But even if it were possible to overcome these and other difficulties when the problem is to study diosmotic exchange, yet it is impossible, in freely suspended cells, to measure the pressure brought about by osmotic action. To render this possible the membranes must be placed against a support, which can offer resistance to ordinary pressure, but which is relatively easily permeable to water and salts. The plant cells furnish us with the model desired for imitation. In these the plasma membrane [*Plasmamembran*][2] which, in its diosmotic properties, is similar to the artificial precipitation membranes, is pressed against the cell wall [*Zellhaut*].

In my first experiments, freely suspended membranes were allowed to increase by osmotic pressure until they finally rested on a support that closed one end of a glass tube. If, finally, with some trouble, this was accomplished, other difficulties appeared, in reference to measurement of pressure, which induced me to adopt another course. The precipitated membrane, even under slight pressure, would be squeezed through the pores even of the thickest linen and silken textures--i.e., the continually growing membrane appeared on the other side of the texture, in different places, in the form of small sacs, which further increased in size and finally burst. Attempts to use thicker material as supports, such as parchment paper or porcelain cells, did not yield favorable results, for reasons which I shall leave undiscussed here.

I obtained the first favorable result by proceeding as follows: I took [*unglazed*] porcelain cells, such as are used for electric batteries and, after suitably closing them, I first injected them carefully with water, and then placed them in a solution of copper sulfate while, either immediately or after a short time, I introduced into the interior a solution of potassium ferrocyanide. The two membrane-formers now penetrate diosmotically the porcelain wall separating them and form, where they meet, a precipitated membrane of copper ferrocyanide. This appears, by virtue of its reddish-brown color, as a very fine line in the white porcelain, which remains colorless at all other places since the membrane, once formed, prevents the substances that formed it from passing through.

These "internal" membranes, deposited in the interior of porcelain walls, I have used, moreover, almost exclusively for preliminary experiments, the investigation proper being carried out with "surface" membranes that were deposited on the inner surface of porcelain cells. All the experiments to be described are carried out with the latter kind of membranes, if not especially stated to be otherwise. To prepare these, the porcelain cells were completely injected with, e.g., a solution of copper sulfate, then quickly rinsed out with water, and a solution of potassium ferrocyanide afterwards added. More minute details as to the preparation of the apparatus will be given later, after this general account.

In Fig. 1, the apparatus ready for use, with the manometer m for measuring the pressure, is shown, at approximately one-half the natural size. The porcelain cell z and the glass pieces v and t inserted in position, are shown in median longitudinal section. The porcelain cells which I used were, on an average, approximately 46mm high, were about

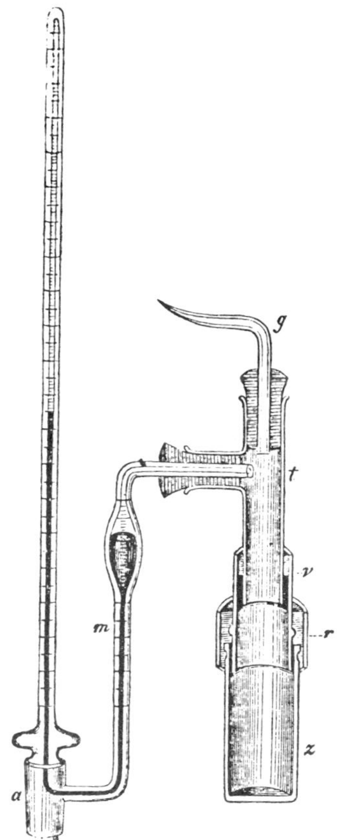

Fig. 1.

16mm internal diameter, and the walls were from 1.25 to 2mm thick. The narrow glass tube v, called the connecting-piece, was fastened into the porcelain cell with fused sealing-wax, and the closing piece t was set into the other end of this tube in the same manner. The shape and purpose of this are shown in the figure. The glass ring r was necessary only in experiments at higher temperature, in which the sealing-wax softened. The ring was then filled with pitch, which also held together firmly the pieces inserted into one another.

If one is careful, it is not difficult to make satisfactory sealing-wax seals even for higher pressures; the greatest care should be taken in joining together the porcelain cell and the connecting-piece. For all experiments at lower and medium temperatures, the narrow annular space between the porcelain cell and the connecting-piece was filled in with two different types of sealing-wax. Most of the space was filled with a good packing sealing-wax, which is harder to melt. Joined to it was a lower ring of softer sealing wax, placed towards the interior of the cell. The softer sealing-wax was produced by adding turpentine or liquid pitch. Only this inner ring, which thanks to its composition did not permit small cracks to form, came in contact with the surface or internal precipitation membrane. The function of the firmer sealing-wax was to prevent the closing-piece from being pushed out by pressure. In order to produce this double seal, the glass tube v is first inserted in the porcelain cell with the less easily melted sealing-wax. The sealing-wax ring, which is extruded in the process as the glass tube is pushed inside, is removed with a suitable instrument. The easily meltable sealing-wax is added and then, when the latter has melted, the glass tube is pressed further in. Greatest care should be taken that there is a very good join between the sealing-wax and the porcelain cell, and that the ring of soft sealing-wax makes a smooth seal toward the inside of the cell. If not, the experiment will fail because there will be leaks. Because the short connecting piece allows one to look inside the cell, I preferred to use it.[3] In any case, it is always safe and easy to melt this wax so that the closing-piece t, when in place, is leakproof.

At temperatures up to 25°C, the seals were completely leakproof at all the pressures that were achieved. At higher temperatures, however, the sealing-wax could soften and allow the pieces that had been cemented together to be pushed apart. For higher temperatures, a cement was used that would not soften, but the seals were made leakproof in the usual manner with sealing-wax. The cement I used, which was suggested by Hirzel (1868:58), is generally very much to be recommended because it is easy to prepare and has many uses. It is produced simply by rubbing together lead monoxide and glycerin. Depending on its use, a heavier or lighter liquid-paste is prepared and, within 24 hours, the mixture turns hard as rock. Incidentally, to make a paste that hardens faster, it is a good idea to add a little water to the concentrated commercial glycerin.

Figure 1 shows how the cement was applied. The glass ring r was placed on a paper annulus positioned around the porcelain cell and then the cement was poured in the annular space. In order to ensure there would be no slipping, I always cut a groove in the porcelain cell, while I blew flanges onto the sides of the connecting-piece, as shown in the figure. Since the porous porcelain cell would withdraw the glycerin from the cement, the cell must first be soaked in glycerin.

In order to ensure that there is a seal between the two glass tubes t and v at higher temperatures, the annular space between the two tubes was filled with sealing-wax in the lower half and with lead monoxide cement in the upper half, as shown in the figure by unequal shading. If the cement is to adhere and to fulfill its purpose, it obviously has to come in contact with the clean glass surface. This condition is so easy to satisfy that I do not feel it is necessary to describe my procedure.

Of course, the lead monoxide cement would not resist acids and alkalis and, if one wanted to work with those substances, it would have to be replaced with another type of cement. The kind of minimal surface reaction produced by copper salts has no practical significance.

All porcelain cells were treated first with dilute potassium hydroxide, then with dilute hydrochloric acid (about 3%) and, after being well washed, were again completely dried before they were closed as already described. Substances which are soluble in these reagents, such as oxides and iron, which under certain conditions can do harm, would thus be removed.

After the apparatus was closed, the precipitated membrane was formed, either in the wall or upon the surface, according to the principle already indicated. In order that this should be done successfully, a number of precautionary measures are necessary, and these will now be discussed. Since I experimented chiefly with membranes of copper ferrocyanide, which were deposited upon the inner surface of porcelain cells, I will fix attention especially upon this case.

The porcelain cells were first completely saturated with water by repeated evacuation under the air-pump, and then placed, for at least some hours , in a solution containing 3% copper sulfate, and the interior was also filled with this solution. The interior only of the porcelain cell was then rinsed out quickly several times with water, well dried as quickly as possible, by means of strips of filter paper, and after the outside had dried off somewhat it was allowed to stand some time in the air until it just felt moist. Then a 3% solution of potassium ferrocyanide was poured into the cell, and the cell immediately reintroduced into the solution of copper sulfate.

After the cell had stood for from twenty-four to forty-eight hours undisturbed, it was completely filled with the solution of potassium ferrocyanide, and closed as shown in Fig. 1. A certain excess of pressure of the contents of the cell now gradually manifested itself, since the solution of potassium ferrocyanide had a greater osmotic pressure than the solution of copper sulfate. After another twenty-four to forty-eight hours the appparatus was again opened, and generally a solution introduced which contained 3% potassium ferrocyanide and 1.5% potassium nitrate by weight, and which showed an excess of osmotic pressure of somewhat more than three atmospheres. If the cell should, moreover, be used for experiments in which a higher pressure was produced, it was also tested at a higher pressure by using a solution richer in potassium nitrate. In these test experiments, of course, any home-made manometer can be used.

We know from experience that in order to produce usable apparatus, it is very important initially to increase the pressure slowly and to maintain this lower pressure for a certain length of time. Of course it is obvious that the membrane, formed without one-sided pressure, can be stretched over small depressions on the inner surface of the porcelain cells, to which it must adhere because of developing pressure. This will certainly occur if it is performed quite gradually, while a more rapid increase in pressure might cause the membrane to tear.

At least one fact seems to speak for the truth of that assumption -- when pressure develops more quickly, the mercury in the manometer at first rises to a certain degree, only to fall again more or less quickly. Soon after this change, smaller or larger reddish-brown copper ferrocyanide stains often appear on the outside surface of the porcelain cell, in support of the fact that the surface membrane has lost its former continuity.

The length of time is also not quite without importance in producing our cells, because the membrane gradually becomes somewhat thicker and more resistant. This can be important, for example, where the membrane closes a pore that permeates the whole porcelain cell, and must counteract the pressure bearing on it by its own resisting ability in such a way that it neither tears nor grows by intussusception; in the latter case, as we know, the membrane would protrude on the exterior surface of the porcelain cell. These and similar considerations make it clear why it is important to displace the air in the porcelain cell, for if the still-growing membrane encounters an air bubble, one of the membrane-formers needed for further growth is lacking at the contact surface with the air bubble.

If the precautionary measures listed above are supplemented by practice, there is a good chance that usable cells can be successfully produced. Toward the end, I had hardly one failure out of 20 cells, while initially I had to struggle against great difficulties and, before I resorted to the expedient of partially drying the cells, was completely unable to produce even one surface membrane. Prior to that, I had worked with internal membranes, which required the same precautionary measures for their production. Normally, I would inject the cells completely with water, then first dip them in a 3% solution of copper sulfate, and only fill the interior with an equally concentrated solution of potassium ferrocyanide after 15 to 20 minutes. Therefore, the membrane is not formed in the middle, but quite close to the interior surface of the porcelain cell. Aside from the fact that this surface membrane has its advantages because it comes into direct contact with the liquid poured in (while in the case of the internal membrane the diffusion processes in the porcelain mass also play a role), the surface membrane can always be produced more dependably. I have even handled a few shipments of porcelain cells in which it was easy to produce a surface membrane, whereas I found it very difficult or even impossible to produce usable internal membranes.

In fact, of ten shipments of porcelain cells, which for the most part came from different factories, only the material of two manufacturers proved to be suitable for producing internal membranes.

The condition of the material is always important, even if a cell can actually be successfully produced. It stands to reason that the most suitable characteristics are provided by porcelain cells which are as porous as possible. Such cells affect the osmotic exchange through the precipitation membrane as little as possible. Best suited to these requirements were the porcelain cells already mentioned above, which I used in my experiments. I obtained these from E. Leybold & Co. in Cologne, purchasing their entire stock. The porcelain cells which this company later procured form the same factory proved to be quite useful for producing surface membranes. However, it was scarcely or not at all possible to use them in making internal membranes, and they were of inferior quality to the cells I had obtained previously. I have found that material purchased from nine other factories does not equal the previously described cells in quality, and so I am unable to quote a source of supply for porcelain cells of the most suitable quality.[4] E. Leybold and Co. in Cologne, by the way, will be able to supply cells, quite adequate for surface membranes, of approximately the same dimensions as those I used.

Eggshaped cells tapering off into a cylindrical neck would have certain advantages, but since the material used in the porcelain cells manufactured for me in this form was inferior as compared to the cells mentioned above, which I found the most useful, I understandably used the latter in my experiments. Here I will note expressly that certain products turned out to be completely unusable for preparing surface precipitation membranes, and remained so even after treatment with acids and alkalis, although the appearance of these cells gave no indication of their unsatisfactory performance. Here, however, their physical, not their chemical property must have been a determining factor, since on the other hand cells made of a chemically different material made it possible to produce suitable precipitation membranes.

In all cases, the development of high osmotic pressure by a dilute solution is already in itself a sure criterion that a precipitation membrane has been successfully formed. If there were flaws in the membrane from the start, or subsequently, this pressure always rose only slightly, and then fell back. In fact, this criterion proved to be adequate in practice, for in all these cases experiments with different cells produced identical results in terms of pressures reached; also such membranes then always turned out to be impermeable to those solutes that are incapable of passing diosmotically through the same type of flawless precipitation membrane.

I could deposit on the porcelain cell membranes of prussiate of iron and calcium phosphate just as easily as membranes of copper ferrocyanide. To produce the prussiate of iron membrane, the cell was first saturated with iron chloride (1.5% solution) and then a 3% potassium ferrocyanide solution was poured inside. For calcium phosphate membranes, a 3% calcium chloride solution and a 6% solution of disodium phosphate ($Na_2HPO_4 + 12 \cdot H_2O$) mixed with a little sodium bicarbonate was used. The calcium phosphate membrane, and membranes made of other suitable substances, such as iron oxide hydrate and iron phosphate, permits us to work with alkaline solutions, by which copper ferrocyanide and prussiate of iron are decomposed. Of course, when using somewhat alkaline solutions, the lacquer seal must be replaced by another kind of seal.

It is probably to be expected that all membranes formed from dissolved crystalloids in solution can also be deposited on porcelain cells. I would also like to think that the formation of such membranes by means of colloidal membrane-formers would probably be successful, even though some attempts to deposit membranes of tannic gelatin glue in porcelain cells failed. However, one should not attribute great significance to this result, since modified methods might very well produce good results, although I had no reason to pursue this matter. Also it might be possible to infiltrate surface membranes with other insoluble substances, a process by which, as Traube (1867:141) showed, the diosmotic properties of the precipitation membranes can be considerably modified. Of course, the same porcelain cell can be put to new use over and over again, provided that the precipitation membrane can be removed. I found it advisable first to remove the sealing lacquer mechanically, as thoroughly as possible, and then to remove the last traces of it by extraction with alcohol. To get rid of the copper ferrocyanide, the cells were then soaked for about 24 hours with dilute aqueous potassium hydroxide to which some sodium potassium tartrate had been added to dissolve the copper oxide; finally, the cells were washed out and treated with dilute hydrochloric acid. I need hardly mention that the prussiate of iron membrane can also be removed by subsequent treatment with potash and acid, while the calcium phosphate membrane can be directly removed with acid. If a lead oxide putty has been placed around the cell, this can of course be removed mechanically, but in order to be sure to remove all the lead it is recommended, for obvious reasons, not to use sulfuric acid or hydrochloric acid, but nitric acid. It is simplest to take apart the connecting piece v and the closing piece t by breaking up the worthless connecting piece with cracking coal, when lead oxide putty has been used.

In addition to porous porcelain, quite a number of other types of material are probably useful for making internal or surface precipitation membranes, but in this respect I have only tested animal membrane [*Thierblase, a membranous sac, e.g., pericardium or bladder*] and parchment paper, and with good results. With parchment paper, I always most easily obtained a diaphragm [*Scheidewand*] whose osmotic behavior was identical with that of a copper ferrocyanide membrane deposited on porcelain, when parchment paper previously saturated with water closed a glass cylinder on one side. Then I submerged the glass cylinder in copper sulfate solution and immediately filled it with potassium ferrocyanide solution.

A completely tight connection between the glass and the parchment paper is easily established if an easily drying spirit lacquer is applied between the two before tying the paper on with thread.

For exact measurements, membranes deposited on porcelain cells are definitely preferable. For, firstly, the parchment paper is stretched by pressure and the flaw resulting from this would not be totally eliminated even if metal sieves were placed under it. Secondly, the precipitation membrane is easily damaged by the stretching and can be counted on to be continuous only where there is an abundance of membrane-formers. Lastly, in all the experiments, after overpressure reached one to two atmospheres, potassium ferrocyanide was pressed through, a process obviously analogous to the pressing of precipitation membranes through closely woven linen. Nevertheless, the membranes deposited in parchment paper are suitable for a number of osmotic experiments, particularly for the demonstration of osmotic pressure. By drawing out a sufficiently wide glass tube into a neck at one end and placing within it a closing piece, t in Fig. 1, in which open manometers can be inserted, we have an apparatus which meets all requirements. If the tube has a diameter of 20 to 25mm, good parchment paper will certainly withstand an overpressure of 3/4 to 1 atmosphere.

The precipitation membranes deposited on porcelain cells form a thin layer on the porcelain material which, when it consists of colored solutes such as prussiate of iron or potassium ferrocyanide, is clearly visible. Broken cells show that this precipitation membrane closely adheres to the uneveness of the porcelain material, and conforms to it by being uneven itself. True, this fact in no way affects the experiments themselves, but it would be of great importance if one had to measure the thickness of the membranes, indeed it would render impossible such measurements, which otherwise could be carried out very accurately optically.

In the case of the surface precipitation membranes in the porcelain cells, I specifically had to determine the numerical value of three quantities under variable circumstances. These were: 1) the movement of water into a cell brought about by an osmotically active substance, 2) the filtration, i.e., the outflow of water under a known pressure, 3) the pressure which results from a state of equilibrium between 1) and 2) caused by an osmotically active substance in a closed cell. The intention of my work was not to determine the ratio of the quantities of water and salt being exchanged, the so-called endosmotic equivalent, for solutes which diosmose through the precipitation membrane.

Let me note here that when I say "osmosis" or "diosmosis", I mean the passage of a substance through any kind of diaphragm. I will no doubt occasionally refer to the osmotic movement of a substance into the interior of a cell as "endosmosis", that is, I will use the word to describe only one particular direction of the osmotic flow. The state of equilibrium between endosmosis and filtration referred to under 3) shall be designated as "osmotic pressure" or "pressure".

It is true that in all procedures it is possible to use the precipitation membranes without the presence of membrane-formers; however, small tears can easily occur, and these may lead to great errors. That is why, unless I had special reasons for acting otherwise, especially when testing the pressure, I added one of the membrane-formers to the inside and outside solution in such quantity that the osmotic counteraction of both solutions was exactly balanced. This was all the more possible, without causing a significant error in the experiments, since experience shows that it takes only very dilute solutions of the membrane-formers (solutions of 0.1% and less) to repair tears that occur in the membrane. We shall come back to this point later.

2. Measurement of Osmotic Water Flow.

To determine the endosmotic movement of water quantitatively, I used the apparatus shown in Fig. 2. The rise of the column of liquid was observed in a calibrated tube s, which had been set in place in the end opening of the closing piece (see Fig. 1) with a rubber seal. The whole cell was immersed in water,[6] whose temperature was determined by accurate thermometers at two different heights.

The relatively low endosmotic increase in volume, and the need to keep an experiment as short as possible, demanded that I choose a narrow measuring tube. The tube I used had a diameter of 1.4090mm, and was divided into millimeters for a distance of 20 centimeters. Exact calibration resulted in a uniform diameter for this divided section, so that at all points the rise of the liquid by 1mm indicated an increase in volume of 1.559 cubic millimeters.

There is no need for a special exaplanation of the adjustment of the column of liquid in the measuring tube, and of the way the apparatus is put together; in passing, let me note that drying out the part of the measuring tube not occupied by the liquid, by means of a wick, is necessary.

Since the volume of the cell is comparatively large --at least 15 cc --a temperature variation of 1°C will become apparent in the measuring tube as a change in the level of the liquid of about 2mm, and it is therefore necessary to control the temperature accurately. The temperature was read to 0.05°C on both thermometers, whose bulbs were located next to the lower and upper ends of the cell; the thermometers themselves were divded into tenths of a degree and exactly compared to a Geissler standard thermometer. During any one experiment, with rare exceptions, great care was taken that the temperature of the liquid oscillated by a maximum of 0.2°C only, and prior to taking a reading the initial temperature was always reestablished accurately to 0.05°C. This was done comparatively easily by touching the glass cylinder with one's hand, or with a cold object. I preferred this easy method to normalizing, by calculation, to the same temperature, which would have necessitated that the expansion value be determined for each cell.

The change in the liquid level of the column in the measuring tube was accurately recorded by means of a cathetometer accurate to 0.1mm, so that any error resulting from this and from temperature differences can amount to no more than a 0.3mm change in height, at most, in the measuring tube. Direct experiments in which osmotic action had been excluded convinced me, moreover, that in such a cae this error is not exceeded by the already mentioned and the yet to be mentioned sources of error, combined.

The difference in the level of the liquids inside and outside the cells is a small pressure of such minute significance for filtration that a further source of error caused by this pressure only needs to be taken into consideration if the endosmotic action is weak. Even then, this error is limited to an infinitely small measurement if

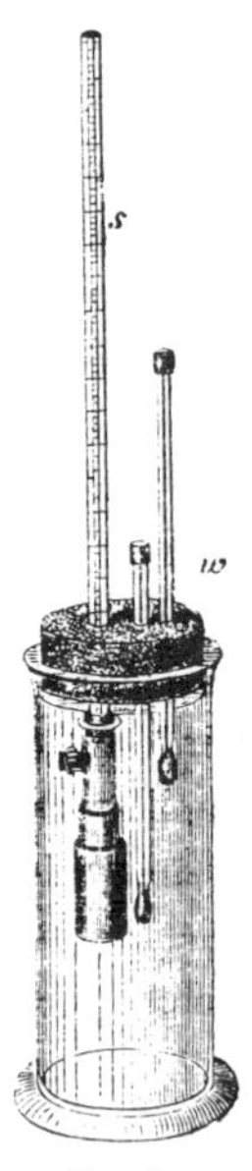

Fig. 2.

the pressure difference is itself a minimum; this can be reached when the liquid level in the measuring tube is set higher than the outside fluid in the glass cylinder corresponding to the capillary rise. For dilute salt solutions, too, this is almost reached when the value of the capillary rise of water is used, which in our measuring tube equals 21 to 21.5mm for temperatures between 8 and 20°C (Buff, 1874:199). The pressure that brings about filtration then amounts to no more than a few millimeters when there is little change in volume in the course of the experiment.

When the stoppers are moved, no measureable error occurs if the apparatus is protected from jolting; similarly, no error occurs as a result of water evaporation in the narrow measuring tube. It is very important to note, however, that the permeability of the membrane decreases with time; through thickening and plugging when the membrane-formers are present, through plugging alone when they are absent. Even in the latter case, such a change cannot be totally prevented, yet when very clear solutions are used, it can be reduced to a minimum. The size of this error can be judged in almost all my experiments, since a series of experiments usually concludes with the same experiment with which it had begun.

Since I did not know the thickness of the membrane, and it was impossible to determine the influence of this and other factors, the endosmotic action under varying conditions could always be studied only comparatively with the same cell. Here it proved sufficient, when changing different solutions, if a cell with a surface membrane was rinsed several times with the new liquid with which it was to be filled.

It was also shown that the state of equilibrium was always established after only 10 minutes, as readings at successive intervals indicated, a procedure which, incidentally, was used in all experiments as a means of control.

The volume increase of the solution inside the cell was too small in all cases to produce any notable variations in concentration. Also, the osmotic passage of the solutes, if it even took place, was so insignificant even for potassium nitrate -- the one substance among those I used which diosmoses most -- that at the end of the experiment it was hard to determine.

3. Filtration under Pressure.

The filtration under known pressure was measured by means of the apparatus shown in Figure 3. A glass tube which was bent as shown, mainly to permit easy handling, connects the cell z with the pear-shaped vessel o containing mercury and water or, as in the cell, a solution of one of the membrane-formers equilibrated with the outside liquid. Into this vessel is set a measuring tube s designed to record the pressure of the mercury column and, set into the base of the vessel, a tube, bent at a right angle, equipped with a glass cock n and a collecting vessel h. After the above parts had been joined together using very tight-fitting rubber stoppers, the first vessel o was filled with mercury for the greater part, then cell z was filled with liquid and closed with a rubber stopper, in the way indicated in Fig. 1, through which a glass tube g ending in an open capillary tube had been run. Finally, the rubber stoppers were securely fixed in place by wire bands.

Now it is important to fill the cell completely with water and thus to displace part of the mercury in vessel o into the collecting vessel h. This is easy to do if fluid is pushed through the capillary with an appropriate device, and if this is repeated until all the air is displaced and, in general, the state I have just described is reached. Now the glass cock n is closed and the capillary tip is sealed off; if water vapor is produced at first, the procedure can be done in such a way that no air is left in the whole apparatus. The apparatus is then, as the figure shows, placed in a bath containing water or the dilute solution of one of the membrane-formers. In order to keep the concentration of the solution unchanged, the bath is covered with glass plates that have two precision thermometers w passing through them. For the rest, the apparatus is held in place by clamping the pressure tube s.

The tip of the right-angled tube above collecting vessel h was connected to a pressure pump, and the mercury in the pressure tube s was driven to the desired height. After the glass cock n is closed, the apparatus permits us to determine filtration velocity under known pressure by means of the downward movement of the mercury column in the calibrated pressure tube.

The measuring tube I used allows the mercury column to rise to a height of 250 centimeters. The tube consists of two parts, joined together and communicating by means of a glass tap. Figure 3(a) and Figure 1(a) illustrate the

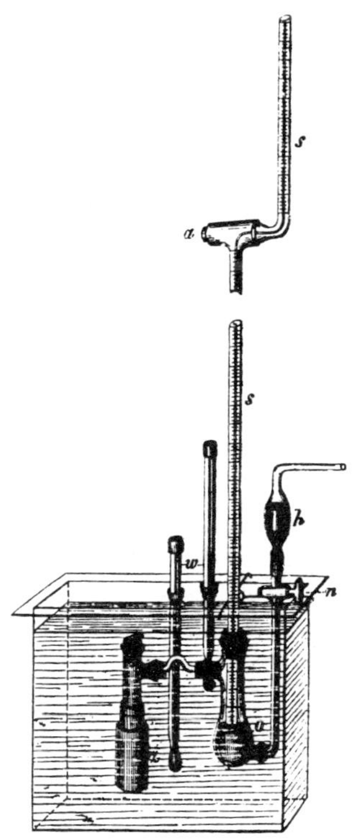

Fig. 3.

capillary through this glass tap and the whole fitting in general, and it is immediately obvious that no pressure will be produced here that will push apart the closing pieces. The division into millimeters continues from the lower tube to the higher; since Geissler's glass taps are so well made, this causes no detectable error as regards pressure. The tightness of the closure has also been completely guaranteed, by Geissler's well-known craftsmanship,[7] for even higher pressures.

But I feel I must mention that both the pressure tube and its connection with the pressure apparatus could be constructed more functionally in a different way.[8] I would have constructed them differently, initially, if I had not had reason to avoid all metal closings since I was planning certain experiments which, as discovered later, proved to be unnecessary.

Since glass tubes of such length are hardly available with a constant diameter, exact calibrations had to be done, at least for the places where readings were taken to determine the amount of filtration. Suffice it to say that the diameter of the lower pressure tube mentioned above was between 1.48 and 1.68mm, and that the upper tube also had a similar lumen.

When establishing the relation between pressure and amount of filtration, the experiment for each individual pressure was continued until the mercury column was lowered by at least 8mm. The accuracy of the observations was controlled by taking readings at a number of intervals. In view of the variable filtration velocity of the membrane, the experiment was performed as quickly as possible and, at the end of a series of experiments, an additional experiment was done at the same pressure with which the series of experiments had begun.

If the temperature of the water in the bath is exactly regulated and if the air temperature, and thus the temperature of the mercury column rising above the water, is kept approximately at the same height, the error that arises from the temperature differences and the reading, together, amounts to no more than a mercury column shift of 0.4mm, though it was probably never as high as that. The elasticity of the rubber stopper produces no significant error, since the change in pressure involved in any one experiment is low (from 10 to 15mm). Of course, every closure must be quite perfect, and the apparatus has to be set up in such a way that it cannot be displaced. That is why I will not mention errors that arise from the apparatus not being perfectly vertical or from the presence of air in the apparatus, and other errors that are avoidable.

To determine the real mean filtration pressure, aside from the magnitude of the pressure developed by the mercury column in the measuring tube, the following factors must be taken into account: the capillary depression of the mercury in the latter, as well as the overpressure of the water contained in the bath and possible fluctuations of the barometer. I was able to avoid normalizing the mercury column to 0°C, since the temperature in each series of experiments was constant and I only wanted to find the relation between pressure and amount of filtration.

4. Measurement of Pressure.

The osmotic pressures were measured chiefly with air-manometers, open manometers being applicable only where smaller pressures were involved. The form of my air-manometer is shown in Fig. 1, at approximately half its natural size. The longer closed limb is connected with the shorter open limb by means of the glass tap already mentioned. An enlargement is blown upon the shorter limb for the reception of mercury. There is a millimeter scale upon both limbs, starting from the same zero point. The scale upon the closed arm is 200mm in length. This arm was selected of small diameter (in the three manometers which I employed it was between 1.166 and 1.198mm), so that the osmotic pressure can be established more quickly, and without any considerable amount of water entering the apparatus. The diameter of the arm, bent twice at right angles, was larger throughout, and was from 7.5 to 8mm in the enlarged space.

Since there was at least a possibility that with longer use some liquid could get to the air closed in the long arm, each manometer was filled with dry air once more after being used five times at the most. Incidentally, a test experiment showed that even after ten uses the highly compressed air still expanded just like dry air when the temperature was raised considerably, i.e., an amount of water large enough to cause noticeable vapor tension had not yet reached the dry air.

Mainly because they would often have to be refilled, the manometers were not made in one piece. At first, a used manometer was taken apart; in the process, it is possible to keep the inside of the long arm clean. Then, after the shorter arm was cleaned, the pieces were put together once more.

When experiments were to be performed at higher temperatures, a fatty substance that remains viscous even under these circumstances must be used to seal the glass cock. By applying a drop of sealing-wax, both arms can be easily fixed in place relative to each other in such a way that they cannot be shifted. I also recommend that a piece of rubber sheet be affixed in such a way that the glass cock is pressed into its position with a certain amount of force.

An apparatus by means of which the manometer can be filled without much effort is shown in Figure 4. The glass tube b is connected with an air pump and the whole apparatus is then evacuated as much as possible. Then air is allowed to enter again very slowly so that it enters the manometer m after being completely dried by sulfuric acid and calcium chloride contained in the potash apparatus. A glass tube d has been inserted in front of the manometer. This tube has a bulge directed downwards which contains mercury. After the air, first rarefied up to a certain point, is dried, the tube and the manometer are tilted in such a way that the mercury is pushed into the manometer by the inflow of air. The level of correct evacuation can then be marked on the simple manometer gauge u for further filling. Obviously it is necessary to repeatedly evacuate the apparatus before one can be sure that the manometer is filled with completely dry air. In use, the space in the open arm of the manometer which was not filled with mercury, was filled with the liquid whose osmotic

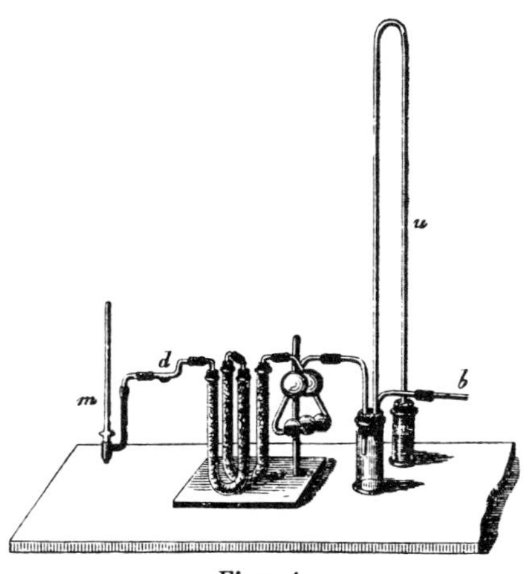

Figur 4.

action was to be tested. The cell was then also filled with this, after the manometer was attached, as shown in Fig. 1, and then the final closing made without leaving any air in the apparatus, in the manner already described with the aid of a glass tube drawn out to a capillary. After the capillary point is sealed off, it is recommended to produce some pressure in the cell, by pushing the glass tube farther in, in order to lessen the time required to reach the final pressure, and at the same time to diminish the amount of water which enters the apparatus. The apparatus can be opened again without any difficulty, after the experiment is over, by blowing open the capillary point in the flame. If the form of the glass tube t allows the rubber corks to expand somewhat at their inner ends, they acquire thus a considerable support. Yet, for higher pressure, they were always secured by tying them down with metal wire (copper or silver wire), as champagne bottles are closed. I have been easily able to close the apparatus tightly so that it would withstand, perfectly, a pressure of seven atmospheres.

The closed cell, as seen in Fig. 5, is fastened to a glass rod passing through a cork, and was introduced into a bath in such a manner that the manometer, as well, was entirely immersed in the liquid. The temperature was measured by the two accurate thermometers. By covering that portion of the bath not closed with corks with a glass plate, evaporation of the liquid was diminished when the bath was filled with a dilute solution of a membrane-former. The apparatus is represented in the figure at approximately one-fourth its natural size. The baths held from 2 to 2.5 liters of liquid.

It is best to place the baths in dishes filled with sand, in order that the manometers may be easily adjusted with accuracy to a vertical position. If the entire apparatus is covered with a bell-jar, and kept in a room of uniform temperature, it is not difficult to keep the thermometers constant for several hours to within less than 0.1 degree. This constancy of temperature is of significance, because the final condition of equilibrium between the osmotic inflow and filtration under pressure is established very slowly, especially at low temperatures and, therefore, before the experiment is completed we are compelled to assure ourselves that the mercury stands at the same height in the manometer for several hours.

In determining osmotic pressures at higher temperatures, the entire vessel was placed in a heating apparatus that was filled with sand and covered with a bell-jar, and whose temperature was accurately regulated. Care must be taken, in passing from lower to higher temperatures, that the junctions of the apparatus are not injured by the increased pressure which is caused by the expansion of liquid and air.

Small osmotic pressures were, indeed, measured with an open manometer, whose longer arm was made of a narrow tube of approximately 0.3mm diameter, in order that the condition of equilibrium might be quickly reached. The form and method of using this manometer requires no special explanation. It should be observed, in passing, that the error of measurement is, in any case, less than 3mm.

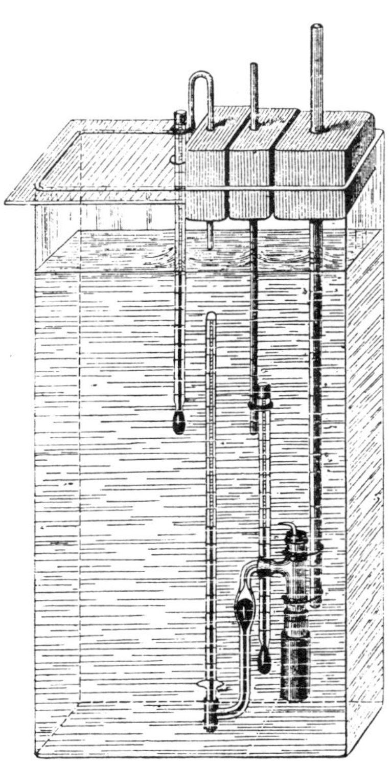

Figur 5

5. Calculation of Pressure.

The calculation of pressure from the readings requires not only the usual corrections, but also a number of corrections dictated by the apparatus and by the way the experiment is conducted. To make calculation as convenient as possible, I proceeded as follows.

At first the manometer, filled only with mercury, was set vertically in a bath filled with water, while the open arm communicated with the atmosphere. The manometer readings, after considering the meniscus error and the calibration table for the closed manometer tube, gave the corrected volume v, which was normalized v° to 0°C and the pressure of a column of one cm of mercury. This normalization requires that we know the following: the temperature of the enclosed air t, the barometric pressure b, the recorded difference in the height of the mercury columns in both arms of the manometer d, and the difference of the capillary depression of the mercury c, which is in favor of the closed arm. Thus we have:

$$v^\circ = v \cdot \frac{(b + d - c)}{(1 + \alpha\ t)}$$

Obviously, this value of v° has to be calculated only once each time the manometer is filled. --The values of $(1 + \alpha\ t)$, or their logarithms, were taken from the tables in Bunsen's gasometric methods.

At first, I obtained the corrected volume V, after reaching osmotic pressure. I normalized it to zero degrees centigrade, V°, and calculated the excess pressure D that results, according to Mariotte's law:

$$D = \frac{v°}{V}(1 \pm \alpha\ t') = \frac{v°}{V°}$$

The osmotic pressure P produced by osmotic action [*P replaces Pfeffer's symbol O throughout this translation*] is now obtained by subtracting from or adding to D: 1) the barometric pressure b', 2) the difference of the mercury levels in the arms of the manometer d', 3) the capillary depression of mercury c, 4) the overpressure expressed as the mercury pressure of the liquid contained in the bath e, which could be equated, without significant error, to the pressure of a column of water whose height extends from the level of the liquid in the bath to the mercury in the collecting vessel of the short arm manometer.[9] Thus

$$P = D - (b' \pm d' - c + e) = D - S$$

Obviously, it is simplest to express the air volumes directly in millimeter divisions of the manometer.

The closed arms of the three manometers I used had dimaeters between 1.166 and 1.198mm and, correspondingly, the volume between two successive millimeter marks is 1.0651 or 1.1247 cubic millimeters respectively. This volume proportion could be regarded as a constant for all sections of these manometer tubes, as the calibration by means of a mercury column indicated. The error produced by the sealed-off tip was also only minimal in all the manometers. In order to obtain the corrected volume, taking into account the mercury meniscus, no correction at all was necessary in one case, while 0.2 or 0.6 respectively has to be added to the recorded volume in the two other manometers to express this volume correction in millimeter divisions for each manometer tube.

The capillary depression for the closed arm of the manometers is calculated to be between 8.1 and 9.6mm. In all the cases, a correction of 9mm could be made without causing significant errors (even when there were temperature differences). To prevent errors caused by adhesion of the mercury, the latter, as well as the manometer tube, had to be kept completely clean. It is generally a good idea to eliminate a possible adhesion, before concluding an experiment, by slightly shaking the apparatus.

A minor error produced by not normalizing the pressure developed by the mercury columns to an equal temperature could be neglected here because it was so insignificant. This becomes more obvious if we remember that the temperature of the mercury in the barometer during all experiments described here was between 12 and 19°C, and that these mercury columns in the manometer are so low that even where there are 20°C fluctuations of temperature, the difference in pressure amounts to less than 1mm. There is also no need to justify the fact that the expansion of glass, the incomplete accuracy of Mariotte's law and a few other insignificant sources of error were disregarded.

The sum of the errors produced by readings, temperature variation and inaccurate adjustment of the mercury column in the manometer will at most equal no more than a 0.4mm shift of the mercury column in the manometer tube. This causes an error of approximately 3mm at an overpressure of one atmosphere and, though this error rises to almost 8mm at four atmospheres, adequate accuracy has been established.

6. Preparation and Check of Experimental Solutions.

I have had occasion to remark that, in order to ensure the continuity of the membrane, very many experiments were conducted in the presence of the membrane-formers, whose concentration was regulated in such a way that both remained in osmotic equilibrium. This was the case with the main membrane-formers I used --potassium ferrocyanide and copper nitrate --when the potassium ferrocyanide solution contains 0.1%, and the copper nitrate solution contains 0.09% by weight.[10] It is my experience that these very dilute solutions are sufficient to repair small damages in the copper ferrocyanide membrane.

In almost all the experiments, the copper solution was used as the outside solution, while potassium ferrocyanide had been added to the solution in the cell that was to be osmotically tested.[11] Because the weighing of copper nitrate is unreliable, a solution of approximately 5% of this salt was prepared, its copper oxide content was exactly determined by evaporating it and making it red-hot, and from this the content of $Cu(NO_3)_2 + 3 \cdot H_2O$ was calculated. A solution of 1% potassium ferrocyanide, made from potassium ferrocyanide dried over sulfuric acid, was also kept on hand, in order to more conveniently prepare solutions of other solutes containing known quantities of potassium ferrocyanide.

Where exact osmotic measurements are involved, naturally, the presence of potassium ferrocyanide excludes the use of substances that combine with or are decomposed by it; however, the sodium salts, for example, could be tested if sodium ferrocyanide were used. It is also important to consider whether, in mixed solutions, an inert substance does not modify the osmotic action of potassium ferrocyanide; furthermore, one must be careful that, for a solution prepared by percent weight, the amount of added potassium ferrocyanide in the unit volume of the solution is proportional to the specific gravity of this solution. On the other hand, the osmotic action is more likely proportional to the amounts by weight in the unit volume of the solution.

Let us first look at the point I have just mentioned. Let the pure solution of a solute contain α amount by weight of this solute per unit weight, and a specific gravity of 1.08. If we are to make a solution which again will contain α amount by weight of the same solute per unit weight, but at the same time contains as much potassium ferrocyanide per unit volume as a pure 0.1% solution of this salt, then (0.1/1.08) = 0.0926g of potassium ferrocyanide and $100 \cdot \alpha$ grams of the solute in question must be dissolved per 100g liquid.

If 0.1g of potassium ferrocyanide had been weighed, it would mean an excess of 0.0074g and, if amounts by weight per unit volume are determining for osmotic action, the pressure of the substance being tested would necessarily be found too high by the effect of those 0.0074grams. I did not examine the osmotic pressure of potassium ferrocyanide, but I have reason to assume that this pressure for a 1% solution is less than a mercury pressure of 150cm. If we assume 150cm, then the pressure corresponding to 0.0074g would equal approximately 1.1cm. Such an error is not large compared to high pressures; solutions of so considerable a specific gravity as the one on which our considerations were based can only be used when prepared from certain colloidal solutes, since crystalloids in such a concentration produce a pressure our apparatus could probably not withstand. Where the specific gravity is 1.02, if we neglected the correction in question, an error of only 0.15cm mercury pressure would result. Incidentally, when preparing all solutions whose specific gravity exceeded 1.015, I always included this correction.

Of all the solutions I tested for their osmotic action, one containing 18% by weight of gum arabic was the most concentrated and also had the highest specific gravity; its specific gravity was determined to be 1.072. Two experiments performed in different cells with the pure 18% solution gave the following pressures for the copper ferrocyanide membrane: a = 118.0, b = 120.4cm mercury, while solutions prepared according to the above principle with 18% by weight of gum arabic gave, for the same cells with a potassium ferrocyanide membrane, the following results: a = 118.9, b = 119.7cm mercury. In other words, the values agreed sufficiently accurately.[12] At the same time this result indicates that the osmotic effectiveness of potassium ferrocyanide is not conspicuously affected by foreign admixtures. What is true for a concentrated and, what is more, a mucilaginous solution can be expected all the more for dilute solutions, that is, providing no decompositions are involved. As a safeguard, I also tested solutions containing 1% sucrose with a copper ferrocyanide membrane, in the presence and absence of the membrane-formers. When the membrane-formers were absent, I found a mercury pressure of 48.9cm; when they were present, mercury pressure was 49.8cm.

Similarly, for two other membranes --prussiate of iron and calcium phosphate --the osmotic equilibrium of the aqueous solutions of membrane-formers was first established. Since, in determining this state of equilibrium for iron chloride, I used a solution of this salt whose iron content was not experimentally determined, I cannot give reliable data in this case. If the actual iron chloride content and that given by the factory are identical in the solution I originally used, then a 0.025% solution of anhydrous iron chloride and a 0.1% potassium ferrocyanide solution, with a prussiate of iron membrane, produce the same osmotic effects. For similar reasons, I am also not in a position to give the exact concentration of the calcium chloride and sodium phosphate, which are in osmotic equilibrium on both sides of a calcium phosphate membrane.

In our experiments, in cases where the solute does not diosmose, the concentration of the osmotically active solutions undergoes only an insignificant change. Basing our results on the manometer whose closed arm had the greatest diameter, a 100mm rise of the mercury column would indicate an uptake of about 0.11cc of water into the cell, whose volume is about 16cc. In order to determine the influence of the elastic compression of the rubber and so on, which must also be taken into consideration, the capillary tip (see Fig. 1) was opened inside a tube in such a way that the amount of liquid ejected could be determined by weighing. Taking into consideration its specific gravity, the amount thus determined was compared with the amount that the manometer observation had showed to have been taken up. Uptake caused by elastic stretching, for a pressure of 2 atmospheres, was 0.05cc, while for a pressure of 4 atmospheres it was 0.09cc. If we note that at the beginning of an experiment, after pushing in the glass tube, which had been drawn out into a capillary, a pressure closely corresponding to those circumstances was produced, we will easily understand that the total uptake of water probably did not even amount to 0.14cc. I am justified in not discussing to what extent temperature variations need to be considered here.

In every case, the specific gravity of the solution poured into the cell and emptied from it after the experiment was accurately determined; this was a necessity for diosmosing solutes to enable us to determine the concentration at the end of the experiment. Assume the initial concentration of the solution used and the decrease in specific gravity during the experiment are known and that the potassium ferrocyanide content of the solution remains virtually unchanged, then the content of osmotically active substance in this solution emptied from the cell can be determined when the concentration of the pure solution (free of potassium ferrocyanide) can be calculated from its known specific gravity. The outside solution in the bath can naturally be regarded as unchanged because of its relatively large volume.

Since even dilute solutions of low specific gravity (e.g., 1.004) can produce high osmotic pressures, the calculations of the specific gravity or the amount of substance in solution had to be sufficiently accurate if the relation of osmotic pressure to the concentration was to be accurately determined. Weighing a glass holding a little over 12g of water can easily be done accurately to the fourth decimal place, but it is not at all reliable if the effect of atmospheric density is not taken into account.[13] Instead of this not always simple calculation, I preferred to determine the specific gravity of the solution in the cell, whose content was known, and that of the solution emptied from the cell at the end of the experiment under the same conditions; this procedure meant a greater number of weighings, to be sure, because no calculation could be regarded as being commensurable with another when air temperature differed by more than 2°C and, at the same time, the barometer showed more than 5mm difference during both weighings. In passing let me note that I always filled the flask with solution at a temperature of 17.5°C, and made sure the temperature of the balance varied by no more than 2°C.

If there had been a difference in the density of the solution that was poured in and of the solution that was emptied out then, strictly speaking, the specific weight of the pure solution of the osmotically tested solute should also have been normalized to the same external conditions. But since this difference was never more than minimal, it was possible to forego doing this when the diosmotic loss was calculated. A degree of accuracy had still been attained for the specific lighter solutions, which corresponds to certainty in the fourth decimal, but not in the fifth.

For example, when a density of 1.004 is assigned to a 1% solution, the calculation of its content from its specific gravity would be accurate to 0.025%. This degree of accuracy is necessary since a presssure of 2.5cm mercury is equivalent to a 0.025% solution, when the osmotic effect of a 1% solution equals the pressure of a mercury column of 100cm --a pressure which is, in fact, exceeded by some solutes, such as potassium nitrate and potassium sulfate.

The determination of the specific gravity is a method that can be generally applied and is at least as accurate as any other method. That is why I did not calculate the content of solutes in solutions by any other method. I did check the purity of the sucrose I used in many experiments, with the help of Wild's polaristrobometer,[14] by determining the rotary power of a solution which contained 10g of sucrose, dried at 110°C, per 100cc. From readings obtained in various experiments, the sucrose content was calculated as being 9.93-10.05% by volume; these values indicate that the sucrose used was pure sucrose. Obviously, this sucrose left only a trace when ashed.

B. Experiments and Conclusions.

7. Structure of the Membrane and Paths of Osmotic Exchange.

In a very significant work which teaches the preparation of precipitation membranes, Traube (1867) showed that these can be produced not only from colloidal, but also from crystalloid components. In order for it to be precipitated in the form of a membrane, a precipitate must meet two conditions, according to Traube (1875:59): 1) it must be amorphous, and 2) the molecular interstices of the membrane must be so narrow that the molecules of the components cannot diosmose through. The latter condition must not be interpreted too strictly, for as Traube already demonstrated, certain precipitation membranes are capable of gradual thickening, which is, of course, incompatible with absolute impenetrability for the membrane-formers. If a small amount of a roughly 3% solution of potassium chromate is dropped into a solution of lead acetate of roughly the same concentration, a membranous precipitate first forms at the contact surface; soon, however, the space enclosed by the membrane becomes clouded as lead chromate is precipitated, and not long afterwards the membrane, which generally shows only slight cohesion, falls apart. It is easy to understand that where there is a more active diosmotic exchange between the membrane-formers, the precipitation membrane cannot exist, and Traube (1867:132) rightly pointed out that because of this exchange not every compound of a substance is suitable for forming a membrane.

Traube's other condition, i.e., that amorphous precipitates are necessary for membranes to be formed, is correct in so far as up to this point we only know of membranes produced from precipitates that appear amorphous. However, it is an open question whether the smallest particles (molecules or compounds of molecules) of the precipitates can have crystalline forms. The observed alteration of polarized light, by membranes of tannic gelatin glue, already observed by Traube (1875:59), indicates there is something to be said for this, but I would not like to base a final conclusion on this alteration. In any case, we must not regard Traube's statement as a law, since it seems to be highly likely that, when membrane-formers are selected correctly, precipitation membranes of unquestionably crystalline minute particles will be obtained, if they have not already been produced. It had not been my intention specifically to study the conditions for membrane formation.

Though our understanding of the experimental results does not necessarily compel us to presuppose a very definite conception of the molecular structure of precipitation membranes, we feel even more an intellectual need for deeper insight because only on the basis of such insight can our thoughts follow the path of a solute through the membrane. Furthermore, only then can hypotheses on molecular constitution in general, and on membranes in particular, produce a multitude of ideas for further research. The following remarks refer primarily to actually described membranes made from colloidal substances, but could also easily be adapted to membranes resulting from precipitation of crystalloids.

In all probability, the finer components of colloids are not the molecules themselves, but compounds of molecules produced by the aggregation of the molecules. As we know, molecules are formed by the reciprocal saturation of chemical affinity units and bonding units of atoms. Compounds of molecules are formed when similar or dissimilar molecules, without rearrangement and disruption of the coherence of the atoms that constitute them, come together into an entity of a higher order. This entity is held together by reciprocal affinities exerted by the molecules as a whole upon each other through forces which result, of course, from the atoms' capacity for acting upon each other, and also from the atoms' spatial arrangement in the molecule. Just as the molecules come together into compounds of molecules, the latter, acting as a unified system, will come together into a whole of an even higher order, and thus will be able to form a larger physical mass. As an illustration of a system which acts as a whole, yet its constituent parts move, let us compare the planets to molecules composed of atoms; then our solar system, held together by central forces, would correspond to a compound of molecules. Just as solar systems in turn exert forces by means of the resultant corresponding to their total mass, compounds of molecules also can be united into a system of a higher order.

Since "compound of molecules" is in itself a clumsy term and is not appropriate as part of a compound word, I propose to call a compound of molecules a "tagma" (from the Greek το ταγμα, a mass arranged according to a law). I believe I can justify such a term all the more as experienced chemists have told me there is need for a precise name for compounds of molecules. Then syntagma is any physical body composed of similar or dissimilar tagmas. Paratagma is a specific term for a body that, predominantly, extends in two-dimensions, such as precipitation membranes. Not until the physiological part of this treatise will it be shown that Nageli's well-founded view on the nature of organic matter is a specific instance of syntagmatic structure. But here I must point out that this ingenious scientist, with admirable perspicacity, inferred a syntagmatic structure for grains of starch and cell walls at a time when chemical science was not yet aware of compounds of molecules as finer constituents of structures (Nägeli, 1858:332). Unfortunately it is not possible to keep the term used by Nägeli, since he called our present molecule an atom, while he called our tagma a molecule.

Compounds of molecules were first introduced in chemistry by Kekule (1861:145,444). The admissibility, indeed the necessity, for such compounds of molecules, aside from direct linking of atoms is, I am sure, recognized without exception today.[15] There is little agreement even in specific cases, especially as views on the valence of atoms vary. Nevertheless, however broadly or narrowly tagmatic grouping of molecules is assumed, some type of cohesion among molecules, for substances containing water of crystallization, seems to be generally accepted. Now, most probably all substances in the colloidal state contain bound water analogous to water of crystallization (this is generally known for a sizable number of definite hydrates, while for others, such as prussiate of iron, copper ferrocyanide, calcium phosphate, etc., studies have demonatrated it), and so we must assume there are tagmas in the water-containing colloids, in which at least water is tagmatically linked to the molecules of the colloid.

The slow hydrodiffusion, and the fact that soluble colloids have little or no capacity to diosmose through membranes that allow crystalloids to pass through very easily, can be understood absolutely only if the finer constituent parts of the colloids (or, more correctly, their range of action) are relatively large particles.[16]

It was this consideration that caused Graham (1862:71) to hypothesize that the colloidal state of a substance may come about when a certain number of crystalloid molecules are arranged together. At all events, however, it is more probable a priori that such a union occurs by linkage of molecules, not atoms. The molecular linkage alone appears reasonable if we note that certain substances, e.g., iron oxide, are found both in a crystalloid and in a colloidal state. It is of little relevance here that we are comparing soluble and insoluble substances; but if we want to go to solutions, it is important to note, for instance, that iron chloride becomes colloidal by combination with iron oxide (Graham, 1862:46). Furthermore, soluble iron hydrate is distinctly colloidal, while soluble ferric salts, whose molecule contains more than three atoms of water and perhaps an acid of very high molecular weight, diosmose very easily through a membrane that allows only traces of ferric hydroxide, or none at all, to pass through. All these considerations indicate that the assumption that tagmas exist and are the finer constituents of colloids, has value as a highly probable hypothesis.[17]

It can be established by agreement whether the term syntagma shall designate a state other than a solid state of tagmatic masses. I myself find it more useful to keep the term syntagma for the solid state, and to use a term like polytagma in other cases.

Not every physical body has to be composed of tagmas, for molecules can (probably) come together directly in an unrestricted way, without previously forming an aggregate, into an entity of higher order. Precisely because of this unrestricted ability for combining, we cannot call a physical body formed in this way, whether it is a crystal or an amorphous body, a tagma. This is because a tagma is only an entity of small dimensions, ordered according to a law from a certain number of molecules, and it goes without saying that it is itself variable with the laws that bring it about. Ultimately, it would be convenient to find a precise term for the composition of unaggregated physical bodies that are formed by the direct combination of molecules. However, I would like to leave it to future scientists to respond to the need and create the name, and to decide whether comolecule or polymolecule (vox hybrida) or a better-sounding word should be chosen.

The absorption of water between the tagmas characteristic of precipitation membranes is, according to our definition, no condition for a syntagma; rather, it is a special case possible in given circumstances. Similarly, any form and composition of a tagma is permissible. I leave it open what form is typical of the tagmas that make up the precipitation membranes. The decided double refraction previously mentioned, which is shown by membranes of tannic gelatin glue under crossed Nicols (prisms), probably speaks in favor of polyhedric tagmas, but is not an entirely compelling argument (Nägeli and Schwendener, 1867:354).

In order for a liquid or solute to be able to diosmose, it must necessarily enter the membrane and, vice versa, when absorption into the membrane takes place, the diosmotic passage of this substance is ensured unless special circumstances successfully counteract it. The following are diosmotic paths: 1) the spaces that remain between the tagmas, and 2) in the case where the substance enters into the constitution of the tagmas, the passage through the tagmas themselves. We will first consider the latter case for water.

Copper ferrocyanide contains water, as perhaps do all colloids; when dried at 100°C, it contains seven equivalents of water, according to Rammelsberg (1863).[18] This, of course, does not exclude the possibility that the precipitate included an additional, more loosely bound quantity of water before drying. As Graham (1862:48) notes, copper ferrocyanide is precipitated from concentrated solutions of the components as an almost colorless gel, which gets its usual reddish-brown color from water absorption only when more water is added. Here, a specific example demonstrates the widespread phenomenon that a salt solution can remove a certain amount of chemically bound water from a water-containing precipitate. However, when a membrane consisting of such a precipitate separates the salt solution and the water, then, leaving other flows out of consideration, there must of necessity be a movement of water through the substance itself, i.e., through the tagmas of our precipitation membrane towards the salt, while an opposing transfer of salt is not obligatory.

The above is restated here in a more general form, based on the continuing state of motion of matter. In a given state of equilibrium (we are still discussing the precipitation membrane and the penetrating water) within unit time, an equal number of water molecules move out of the surrounding medium into the sphere of action of a tagma, as move from that sphere into the surrounding medium. That state of equilibrium is disturbed as soon as, for any reason at all, the ratio of this exchange is altered. This is, for instance, the case when the membrane separates water and a solution of a substance which, due to molecular attraction, increases the number of water molecules passing in the direction from the sphere of action of the tagmas to the solution. This, of course, produces a unilateral water flow in the direction from the water to the solution. Similarly, a water flow must move through the tagmas from the cell into the surrounding water if the water that is in the cell is put under pressure and thereby the kinetic energy [*lebendige Kraft*] of the molecules is thus increased.

Obviously, in all cases, the extent of water flow produced by the difference of molecular movements depends (all other things being equal) on the strength of the chemical binding of the water. The number of water particles that escape the sphere of action of a tagma or a molecule per unit time, due to the particles' state of motion, depends on the chemical binding. The behavior of the copper ferrocyanide described above shows that in copper ferrocyanide at least a certain part of the water must not be bound very strongly. The varying data on the water content of other colloidal precipitates, such as calcium phosphate and prussiate of iron,[19] are perhaps due to the fact that when the drying processes are different, the equivalents of water that escape are not equal. It is impossible to judge, with the data we have at present, whether in the colloidal precipitate certain quantities of water are so loosely bound that they are always separated during the normal drying procedures. I have undertaken no research of my own in this direction.

In analogous fashion, other substances can probably move (diosmose) through tagmas, molecules, etc., as soon as they enter the constitution of the latter which, in certain circumstances, obviously may be considerably altered by the process. I feel that I can only speak of constitution here, since we may probably regard each time a substance enters into the molecular aggregate of a tagma as a case of chemical binding. The amount of the substance thus being bound, and the nature of this binding, are of course an important co-determining factor for the amount of this type of diosmotic exchange.

The diosmotic exchange of gases through layers of liquid is basically analogous to the passage of liquid and dissolved substances, to the extent that the gas is absorbed on one side of the layer of liquid and re-enters the gaseous state on the other side. Here, the gas movement is toward the side on which the gas is at the lowest partial pressure.

There is an additional path for a substance to pass through a membrane offered by the spaces existing between the tagmas, whose existence is required by the discontinuity of matter. True, these intertagmatic spaces might be of minimal size and not even allow water to enter; this would be the case for every syntagma impermeable to this medium. However, in our precipitation membranes, the water certainly enters these spaces. This follows immediately from the fact that even solutes which are proven to be indifferent to the precipitate itself, i.e., which cannot pass through the tagmas themselves, pass through the membrane diosmotically. As long as this latter path is not ruled out, however, we naturally cannot directly conclude, from the fact that a membrane is permeable for a certain substance, that the movement takes place exclusively in intertagmatic spaces. Using the criterion mentioned above, it can be proved for all precipitation membranes I have encountered that there is water uptake into these spaces. Moreover, where there is permeability, water uptake is necessary because the membranes never consist of only a single layer of tagmas. But in order to be able to transfer water to each other, tagmas arranged in the direction of the thickness of the membrane, which do not touch directly, require that the intertagmatic spaces themselves contain water.

The fact that even certain solutes that normally diosmose easily are incapable of passing through precipitation membranes can leave no doubt that intertagmatic spaces are minute; however, the question is whether the entire extent of these spaces is controlled by forces of attraction exerted by tagmas.

As Poiseuille inferred and Wilhelmy (1864:12) showed experimentally, a condensed layer is formed on the surface of a solid body immersed in adhering liquid.[20] This layer is variable as a function of the distance from the surface of the body. Such a layer must probably also form around a tagma, provided the liquid adheres, and there can be no doubt of that for precipitates that are capable of being wetted. Plateau tried to calculate the radius of the sphere of action for the force of attraction extending from a solid body. Quincke (1869:402) did so directly, and both determined that it was on the average approximately 55 nanometers. Taken absolutely, this quantity is very small; however, compared to molecules and molecular interstices, it would be large, even if it were still smaller.

Even where there is a small condensation in the variable layer, powerful molecular forces must have been active, since it is difficult to condense liquids. There is also no doubt that the mobility of the water particles in the altered layer is very much reduced. In particular, mass flow will not immediately be produced by pressure. As Poiseuille (1843:50) established, this is also true for capillaries where a stationary wall-layer of the liquid medium forms the canal in which the liquid flows. Of course, this condensed wall-layer is not completely stationary; even in this layer a pressure difference must cause a movement of liquid in the direction of least resistance. Within certain pressure limits, however, what will be produced will be not a mass flow, but a movement as in any liquid bound by molecular forces, i.e., a molecular movement.

Accordingly, we wish to make a distinction, which is true for capillary spaces also, between mass flow as a "capillary movement of water" and between "molecular movement of water." By the latter we generally mean water movement taking place in spaces which are within range of the sphere of action of solid bodies. This movement will still occur when the diameter of the capillary space falls below the diameter of the sphere of action.[21]

Within the field of molecular forces, the passage of water takes place through the tagmas themselves, and I shall also include this path within the category of "molecular movement of water." For, after all, the action extending from the tagmas as a unified whole is due to molecular forces. The passage of water through and into the tagmas can, where necessary, be further differentiated as diatagmatic and amphitagmatic. It is of no importance to us whether the water taken up into the tagmas is to be regarded as chemically bound, or whether the water layered around the tagmas is to be regarded as physically bound. Without arbitrarily drawn boundaries, that question cannot be decided at all. What I just said about the passage of water is essentially also true for all liquids or solutes which enter the field of molecular forces.

There is no doubt that in an animal membrane, and even more so in porcelain cells, there are spaces that allow the capillary movement of water; such spaces probably occur in all diaphragms previously used in osmotic experiments, including the cell walls of plants. In principle, insofar as any substance is able to penetrate the finer components of such a membrane --be they tagmas, molecules or other particles --the three previously differentiated paths will be there jointly in the same membrane. It is very probable that this is true, but difficult to resolve without specific research. Nägeli's penetrating research compels us to assume the vegetable cell walls and organized structures in general are composed of tagmas; but of course it does not necessarily follow from his research that there are three possible paths. Indeed, Nägeli considers the tagmas themselves to be incapable of imbibition; however, such a view is untenable in this general form. Perhaps it is not superfluous to state that well-known facts demand various possible pathways for the diosmosis of a substance, no matter what one's views on molecular constitution may be, regardless of whether the membrane is composed of molecules, tagmas, or other particles. I probably do not need to justify why I preferred to base my thinking in terms of the likely structure of precipitation membranes.

In accordance with what has just been stated, a boundary layer must always form at the wall of the pores, in which layer the solution is differently composed than on the axis of the pore. This occurs as soon as the attracting or repelling forces extending from the wall bring about a specific distribution of the solutes and the solvent in the boundary layer, through an unequal action of the forces on the solutes and solvent. Brücke's (1843) theory of osmosis is based on the existence of such a boundary layer in pores. When water is attracted with greater intensity by the field of the forces extending from the wall then, obviously, a more dilute solution must be absorbed into the membrane from a salt solution. In fact, this was shown to be true for the animal membrane by Ludwig's (1849:15) experiments,[22] when he studied the salt content of the liquid absorbed when the membrane was immersed in a salt solution of known concentration.[23] True, these experiments do not strictly argue for a variable wall-layer in the pore, for the same results must also be obtained when water, or more dilute solution, penetrates the interior of the finer particles, as for instance in tagmen. No one has yet shown that such uptake does not occur for the animal membrane. Obviously, this boundary layer need not be denser than the adjoining liquid, though the boundary layer was formed by condensation. Thus Wilhelmy's (1864:13) experiments showed that a wall-layer of lower specific weight was formed when plates were dipped in fairly concentrated glycerin, clearly, because this layer consisted of a more dilute solution. This wall-layer becomes a more dilute solution when the forces extending from the wall act in the corresponding way upon the solvent and the solute.

It was Fick (1855:86) who first expressed the idea that Brücke's above-mentioned theory of osmosis does not adequately explain all the phenomena. Fick (1857:296 and 1866:30) further differentiated between substances penetrating into capillary spaces and those penetrating the substance of the membrane itself. Accordingly, he contrasted "pore diffusion," an exchange of water and salt that takes place in capillary spaces, with "endosmosis," an exchange through the substance of the membrane. Granting that the views on which this differentiation is based are correct, I still believe it is necessary for practical reasons to describe every exchange, through any type of membrane,[24] as "osmosis" or "diosmosis." Take an animal membrane, for instance: as Fick also assumed, pore diffusion and endosmosis (as Fick understood them) definitely occur simultaneously; very often, even in precipitation membranes, it is impossible to determine definitely whether capillary passage is entirely absent.[25] Indeed, it is conceivable that where only molecular passage of a substance occurs in a set of given circumstances, capillary osmosis will also begin to take place as external conditions (such as temperature) change. Moreover, capillary osmosis is not possible without simultaneous molecular osmosis whenever there is a condensed wall layer. If we keep the general terms osmosis and diosmosis, it is possible to designate the mechanism of passage more specifically with the terms "capillary" and "molecular" osmosis. If an even more specialized term is necessary to designate the path travelled by a substance, the expressions diatagmatic and amphitagmatic and possibly intramolecular and extramolecular are available. As I said earlier, I shall probably also call the osmotic flow moving into the interior of a cell endosmosis.

The word diffusion (hydrodiffusion) shall be used for the intermixing of solvents and solutes, as caused by the molecular forces characteristic of them; the term probably corresponds to what is current usage. Such a true diffusion will obviously also be possible in pores of a certain diameter; in questionable cases it is preferable to speak of capillary diffusion. However, intermixing of substances influenced by molecular forces within a diaphragm should be called molecular diffusion, not simply diffusion, except where the context of the sentence also encompasses the effect of other forces.

Each substance, dissolved in water or in another solvent, that is not taken up into the tagmas themselves can diosmose only through the intertagmatic spaces, and of course can do so only if the size relationship between the solute and the space accessible to it permits this. This space may be of different size for different solutes in the same membrane. The two most extreme cases would be 1) if the molecule (or tagma) of a solute does not penetrate at all the sphere of action of the tagmas (or other finer constituent parts of a substance), and 2) if this whole sphere of action is accessible to a molecule. The distribution of solute and solvent in the sphere of action of the tagmas is at all times dependent on the interaction between the tagmas on the one hand, and on the interaction between the solvent and the solute on the other hand. It depends on these active attracting and repelling forces; that is why not only the above extremes, but all the intermediary cases, as well, are possible. Once this has been noted, it is clear that the permeation of a substance does not depend exclusively on the diameter of the intertagmatic spaces. The negative or positive result of diosmotic experiments with different substances and a membrane or, vice versa, with different membranes and the same substance, does not yield a value for the relative sizes of the dissolved molecules, as Traube (1867:141) assumes. For different substances, the very same membrane can be like a sieve with holes of different sizes.

The determination of the relative sizes of the dissolved molecules requires that we know the distribution of solute and solvent in the sphere of action of the tagmas as a function of distance from the tagma. However, this function is determined by several variables. For one thing, it depends not only on the interaction between membrane particles and solute, but also on the forces acting between the solute and water, and the kinetic energy of the molecules. At this time, there is little hope that we can even come close to discovering this complex function.[26]

The difference between the solute diameter and the area of the intertagmatic spaces accessible to a solute can, at most, equal twice the radius of action of the tagmas. Even if this difference must keep getting smaller, it is not an infinitesimal quantity compared to the dimensions of the molecules. This is understandable and also becomes evident if we compare the determinations of radius of action and molecular size, of course the latter cannot claim to be at all accurate. As I stated before (p. 38), Plateau and Quincke established the radius of the sphere of action for immersed plates. They determined that the range over which the liquid's cohesive and adhesive forces are influenced by the plate is about 55 nanometers. Taking into account several considerations, Thomson (1871:66) came to the conclusion that the mean distance between the centers of adjacent molecules in liquids and transparent solids--i.e., more or less the diameter of a molecule, including its sphere of action --must be less than 0.1 nanometer and greater than 0.005 nanometer. Incidentally, it should be noted, though I do not wish to contradict Thomson's conclusions, that the diameter of the tagmas of solutes can be larger.

According to the preceding, a solute will probably find it impossible to pass through a precipitation membrane if its molecular size does not permit passage through that intertagmatic space that lies outside the sphere of action of the tagmas. Since a number of crystalloids do not diosmose either, however, the space which permits capillary movement of water (as defined earlier) must have at least a very small diameter. Now if we consider that the whole sphere of action accessible to water could scarcely be inacessible to a solute, it is very likely here that all the intertagmatic spaces are located in the field of the sphere of action of the membrane particles, so that only molecular osmosis takes place in such precipitation membranes. Obviously, the diosmotic passage of a substance is not influenced in the same way in all regions of the sphere of action.

Other considerations also --for instance, those based on filtration velocity under pressure, and on the molecular forces required for the cohesion of the tagmas --would not bring us further than to a tentative agreement in answer to the question whether only molecular movement is possible in a membrane. The likeliest probability seems to be that the mean distances between tagmas, at least as measured in each spatial direction, are equal. Still we must not lose sight of the possibility that ring-shaped or other suitable arrangements might permit interstices of very different widths. We could draw some conclusions, at least as regards the possible mean size of interstices, on the basis of the amount of water taken up or on swelling phenomena, and so on. The argumentation might be similar to that so penetratingly used by Nägeli (1858:333,351) to discover the molecular structure of starch grains. Depending on the result of such considerations, we might be able to answer the question whether tagmas possibly have a spherical shape, or whether a polyhedric form is a necessary requirement. I did not include in my study this and other points that were really unrelated to the actual goal of my research, and so I shall let it go at this short reference, and leave the related questions untouched.

Consider, contrary to probability, a precipitation membrane that is found to have spaces permitting capillary osmosis, at all events the possible diameter of these spaces is very small. It is hardly conceivable that any molecule could pass through a measurably thick membrane without thus entering the sphere of action of the tagmas, which doubtless form many layers. This becomes even more understandable when we consider, unless conditions change, the way new tagmas are formed, gradually thickening the membranes. This process does not permit an arrangement (in rows) of the tagmas to form capillary spaces running in straight lines transverse to the membrane surface, and thus a solute has to travel a tortuous path.[27] Thus, in general, in our precipitation membranes only molecular osmosis will take place. While in animal membranes, cellulose membranes and diaphragms that act similarly, both molecular and capillary osmosis will take place simultaneously.

According to Exner's (1874) studies, a combination of molecular and capillary osmosis also takes place during the exchange of gases through liquid lamellae. For the same lamella, the ratio of the amounts of different gases passing through (without pressure) is expressed by $c \div \sqrt{d}$. That means the osmosis of a gas is directly proportional to the absorption coefficient (c) and inversely proportional to the square root of the density (d). These experimental results force us to assume that a lamella produced from water mixed with a little soap still contains pores. Part of the gas would flow through these pores according to Graham's law, passing as though through a narrow canal in a thin wall.[28]

When a precipitation membrane is prepared, it is always in a condition where it has imbibed water. Of course this water of imbibition can be partially removed by means of substances that attract water, and can be taken up again when this substance is removed. Like the animal membrane and the cell wall of plants, the precipitation membrane is a body capable of swelling, whose volume can fluctuate both by the simple discharge of water from the intertagmatic spaces, and by the discharge of water from the tagmas themselves. I did not try to decide the question whether the membrane can return to its earlier state after total dessication. However, I did discover, by measuring the intensity of the osmotic water flow, that the stationary state corresponding to the conditions is always quickly reached when a concentrated solution in the cell is replaced by a dilute solution. Perhaps it is not superflous to point this out since, according to Fick (1857:315), the salt flow going through a collodion membrane at first increases in intensity, and always seems to take a longer period to become constant. Even very dense animal membranes, such as the cornea of an ox, reach the stationary state relatively quickly, according to Eckhard (1866:98), when they were dried before use.

There is no doubt that both sides of a precipitation membrane are completely identical as regards osmotic action. Where membranes are not homogeneous, it is obviously not a matter of indifference whether the outer or inner membrane surface is turned toward the salt solution.[29]

8. Diosmosis of Solutes.

While the previous section basically discussed the mechanism of imbibition by membranes with respect to osmosis, this section will examine specific osmotic issues.

The simplest case of osmosis exists not during reciprocal exchange but when only unilateral movement of a substance, such as water, occurs through a membrane due to osmotic action, a process which can be repeatedly reproduced with Traube's precipitation membranes. Except for Traube, osmotic investigations were almost exclusively directed to finding the ratio in which salt and water exchange.[30] This ratio, expressed by the quotient of water to salt, as we know, was called the endosmotic equivalent by Jolly. Such an equivalent is out of the question if salt does not pass through the membrane at all, i.e., if the equivalent were to become infinite. Also, for physiological questions, a knowledge of the endosmotic equivalent has little importance. Of course, if we wish to learn more about certain osmotic processes, and study and analyze the forces which cause and control the osmosis, the ratio of the quantities being exchanged will always be significant.

The diosmotic exchange of solutes is frequently encountered in tissues and, especially, in plant cells, where the volume remains constant. Vice versa, there is often water flow without the outward movement of solutes whereby the volume changes. Obviously, various combinations frequently occur. An exchange, which proceeds according to the rules of the endosmotic equivalent (as this is determined in the experiment without pressure difference), is not necessary for the migration of substances in the living plant cell. Also, this type of exchange is not necessary for other phenomena that depend on osmosis, and this exchange is never realized in the living plant cell as long as the internal pressure regulates the intensity of the water flow.

Physiology addresses certain urgent questions to us: 1) can a substance diosmose through a given membrane at all and, if so, in what quantities under the given conditions and 2) what final pressure is produced, by a flow directed into a cell, during a state of equilibrium between this osmotic flow and the amount of water being filtered out under pressure? I addressed this second question using precipitation membranes to seek the cause for the often quite high hydrostatic pressures in plant cells. I did not attempt specifically to answer the second question, and was justified in doing so from the physiological standpoint, since the data already available seem adequate, for now, and can serve as lodestar in studying the diosmosis of substances through such membranes, which are important for the uptake of substances in plant cells.

I just mentioned some of the reasons that prevented me from doing specific research on the diosmosis of solutes and, perhaps, determining the endosmotic equivalent. I refer the reader to Traube's (1867:134), experiments studying diosmosis in a series of substances; I shall merely tell you a few of my observations.

I found membranes of copper ferrocyanide absolutely impermeable to gum arabic and dextrin (pure). After 15% and 20% solutions of these substances had remained in the cells up to 14 days, no trace of a reaction could be obtained with Fehling's solution in the outside liquid after this had been concentrated by evaporation to as small a volume as possible, and had been previously boiled after adding some sulfuric acid. Since I left the cells untouched when membrane-formers were present, the test of the liquid afterward obviously had to be adapted to the prevailing conditions.

When a sucrose solution of up to 5% concentration was used and allowed to stand for 12 days at temperatures of 12° and 20°C, no transfer of sucrose could be shown. However, when sucrose solutions of a concentration of 10% and more had been brought into the cell, a minute amount of copper suboxide was obtained every time from the evaporated outside liquid. So the copper ferrocyanide precipitation membrane allows sucrose to diosmose, though in very minute quantities.[31] Similar results were obtained with a membrane of calcium phosphate.

Potassium chloride and potassium nitrate pass through the copper ferrocyanide membrane more easily. I could give a few quantitative data for potassium nitrate, data that can be derived from the change in specific gravity of the solution inside (the cell) whose volume has at the same time remained constant. I will not do so, since the unkown thickness of the membrane influences the amounts of salt passing through in equal periods of time. The copper ferrocyanide membrane is also permeable to ammonium acetate and to dilute hydrochloric acid.

Similarly, I found that after six hours small, but definitely demonstrable, amounts of a 5% solution of potassium sulfate had passed into the outside liquid. Incidentally, this result is consistent with Traube's (1867:139) statement that the copper ferrocyanide membrane is impermeable to that salt, if we remember that Traube worked with membranes whose area was smaller and with more dilute solutions; also, the time periods of his experiments were shorter.

We shall have to admit one thing as a result of this --new experiments will be needed to decide whether a membrane may be regarded as completely impermeable. True, osmotic experiments and the conclusions based upon them are hardly affected by the passage of minimal quantities of salt, but even a very slight degree of osmosis may become significant for physiological processes.

9. Osmotic Water Flow without Diosomosis of the Osmotically Active Solute.

Endosmotic water flow takes place most simply when the osmotically active solute does not pass through the membrane. Then a suction develops on the inner membrane surface due to the molecular forces which, together, are the basis of osmosis itself. This suction causes a flow of water through the membrane, and this flow obviously has to overcome the same forces of resistance as any movement of water produced in any other way.

During unit time, for a membrane in contact with water on both sides, an equal number of water molecules fly from the surrounding water into the range of the sphere of action of the membrane particles and, vice versa, from that range back into the water. However, if one side of the membrane now comes in contact with the solution of a solute that does not diosmose, the molecules of this solute bounce off the membrane. Since unit surface area per unit time is now hit by a smaller number of water molecules than before, while (all other things being equal) a basically equal number of water molecules still escapes from the sphere of action of the membrane particles, then a water flow moving into the salt solution must necessarily result. I shall later describe how a water flow thus produced is always only a part, and most commonly only a small part, of the water movement that occurs when salt molecules that have approached the membrane exert attraction on the water particles which lie on the boundaries and within the range of action of the membrane particles.

It is significant that the molecular forces acting between membrane particles and salt molecules must necessarily intervene. As previous considerations show (p. 38), the molecular forces will cause redistribution of the solute and water at the boundary layer that exists on the inner surface of the membrane. This distribution corresponds to the affinity --excuse the expression --between water, salt and membrane particles. The nature of the distribution may be different from that of the adjoining solution.

Such a layer, in which water and salt are distributed as a function of their distance from the membrane particles, must extend over the whole membrane surface whenever a substance does not diosmose. In what follows, I shall refer to this layer as a "diffusion zone." The nature of this diffusion zone is always an important factor for the intensity of the water flow into the cell. For it is quite clear that this water flow increases with the concentration difference in adjacent lamina of the solution.

The affinity between salt and membrane will not affect the osmotic water flow directly unless special effects (e.g. chemical effects) are involved, but it will regulate it by virtue of the nature of the diffusion zone. The driving force for the water flow is mainly the attraction between salt and water molecules, and is also determined by a factor I have already mentioned, i.e., the unequal ratio between the water molecules that impact on unit surface of the membrane, and the water molecules that come flying out of it. These, however, are the same molecular forces that cause the intermingling of salt and water molecules in free hydrodiffusion; for in the zone, under special circumstances, diffusion of a special kind is the driving force of the osmotic water flow. When water molecules are removed from the membrane surface, a flow of water then obviously moves through the membrane as a consequence.

The molecular forces which determine the nature of the diffusion zone also act to maintain it in a stationary condition. Thus, necessarily, in the range of action of these molecular forces, there is regulation of the movement, which aims at changing the diffusion zone, of the salt molecules. Similarly, these molecular forces attempt to prevent the expansion of the diffusion zone. The osmotic water flow that has been produced attempts to expand the diffusion zone by repelling salt particles from the membrane. However, the active force of this water flow, a flow that is only moderate, is obviously only minimal compared with the molecular forces that work to maintain the stationary condition of the diffusion zone. Furthermore, the molecular forces of attraction increase rapidly with decreasing distance between two molecules (or other particles of mass). Thus, this inflow will not be able to cause the diffusion zone to become noticeably larger. Indeed, this is in accordance with experiments I shall discuss later on the relationship existing between water flows observed without pressure, and the interrelationship between the corresponding final pressures. Once this pressure has been reached in a cell the previous cause for extending the diffusion zone disappears, for now the ingoing and the outgoing water flow are equal. Now, since the water flows created by different solutions of non-diosmosing solutes are in the same relative relationship as the pressures produced by the same solutions, the kinetic energy of the unilateral water flow cannot significantly enlarge the diffusion zone. Let me note here that the resistances that have to be overcome by water flow in a precipitation membrane increase proportionally to the intensity of this water flow.

When we attribute osmotic driving force to a diffusion process taking place under special conditions (unless perhaps chemical actions are involved), there is at least a prospect that we shall be able to apply laws obtained for free hydrodiffusion for osmotic processes as well, and primarily for osmotic processes where the active solute does not diosmose. True, these diffusion laws themselves cannot yet be related to the molecular actions underlying them. Of course anyone who tries to throw light on osmotic processes in this way still faces enormous difficulties. A simple comparison between the work produced by the water and salt particles during hydrodiffusion and the osmotic inflow obviously cannot produce a result without further consideration. This is so even when, at the same time, we obtain a value for the suction force actually developed by calculating the resistances that have to be overcome by the water flow in the membrane. The diffusion zone for various substances may not only be different in extent, but may also behave quite differently as regards the composition of zones that correspond spatially, and as regards the change of this composition with distance from the membrane particles. There is still no basis for making a decision on these points. Under the condition that within the region of the action sphere controlled by the membrane particles a solution of a defined content exists, we are also unable to judge in individual cases whether there is a sudden, more or less abrupt, transition to the concentration of the solution contained in the cell. In general, the comparison alluded to above might reveal reasons for deciding if the solution in the diffusion zone is more dilute or more concentrated than the adjacent liquid. For reasons stated earlier, we would generally expect it to be more dilute if a solute that is indifferent to the membrane subtance is involved. There is also the possibility that when the affinity between the membrane substance and water is quite dominant, the membrane substance is initially covered by a layer of pure water.

According to the above we should not expect, indeed it would be pure chance, that the velocity of the water movement during hydrodiffusion and during osmotic inflow for different solutes should be in the same ratio. The ratio of the water movement during hydrodiffusion is given by the diffusion constants of different salts, reduced to equal units, since a volume of water proportional to these constants is flowing in the opposite direction. According to Beilstein and Voit's (1867:233,433) experiments,[32] we obtain the following values if the constant for sucrose is set to equal 1 --for potassium sulfate = 2.24, for potassium nitrate = 2.9. For one percent solutions of the above substances,[33] mean values were obtained for the osmotic water flow in copper ferrocyanide from two comparative studies, each done with a different cell (again taking the flow produced by sucrose as the reference unit) are as follows: potassium sulfate = 4.39, potassium nitrate = 4.61 (section 17, Table V).[34] Even without calculating the water volumes corresponding to these diffusion constants, we can still see that the diffusion constants are not in the same ratio as the osmotic water flows.

Though the diffusion velocity of a solute gives no definite information on its osmotic effect, after what was said earlier we can generally expect that a substance that diffuses slowly will also produce a smaller osmotic water flow, for the diffusion velocity is itself dependent on the molecular forces acting between water and salt. This expectation is fully confirmed by experiment. With all colloids, known to diffuse only slowly, there were only low osmotic pressures. These colloids were not diosmosing substances and the low osmotic pressures observed also give, relatively accurately, the ratio of the osmotic water flow that takes place without pressure relative to that for sucrose. Since we shall return to this point, let me simply mention that, for instance, a 6% solution of liquid gelatin glue gave an osmotic pressure of 24cm, and a 6% solution of gum arabic gave 24 to 27cm mercury.[35] A sucrose solution with the same concentration would have raised a mercury column approximately 290cm (see Table 8 in section 14, "Osmotic Pressure"). The liquid gelatin glue,[36] and the gum arabic, contained a considerable quantity of crystalloid

10. Osmotic Water Flow with Diosmosis of the Osmotically Active Solute.

Before we speak about osmosis and, specifically, the water flow when, at the same time, the solute diosmoses, I will remind the reader that capillary or molecular osmosis are possibilities and that, in the latter, the path may lead around and through the mass particles of the membrane. First, let us look at the case on which Brücke based his osmotic theory: here there is a narrow pore filled with liquid in which a core cylinder is located outside the range of the forces extending from the wall. The consequences of this interpretation were developed in detail by Fick (1855:74) and therefore I can limit discussion here to essentials.

In the core cylinder involved here, a solution is imbibed unaltered. Also, a boundary layer whose composition depends on molecular forces, and changes with distance from the membrane, is formed. In very many cases this boundary layer most probably decreases in concentration the closer it is to the membrane, and likely becomes pure water directly next to the membrane. If one opening of the pore adjoins a quantity of water which is to be regarded as infinitely large while the other opens into a salt solution then, due to (molecular and capillary) diffusion processes, the concentration in the pore itself must increase from zero to the specific gravity of the salt solution, or possibly to the maximum concentration that can exist in the individual concentric annuli of the boundary layer. In the core cylinder centered on the capillary axis and in all concentric annuli in which a salt solution can exist at the same concentration as that inside the cell, there is no reason for the cell contents to increase in volume, since salt and water are exchanged in equal volumes (as also happens in hydrodiffusion).[40] But if the concentration in a concentric annulus of the boundary layer is lower than that of the salt solution, a diffusion zone has to form at the boundary between them. The diffusion zone produces water flow into the cell, and this is greater than the salt flow leaving the cell. The intensity of this unilateral water flow depends on the difference in concentration between the concentric annuli and the salt solution, the width of the diffusion zone, and the nature of the solute. Here, too, the formation of the diffusion zone (for reasons already discussed above) must be caused by molecular forces acting between salt, water, and membrane particles. The form of the diffusion zone will be only minimally affected by the water flow produced, and will not be much dependent on it, as Fick (1855:77) assumes.

Thus in certain zones of the boundary layer a unilateral water movement is brought about by osmotic action, causing the contents of the cell to increase in volume. If, as a result of this, the pressure in the closed cell rises, then the entire pore allows the outward-directed filtration flow. It is in the core cylinder, which is ineffective as regards the volume increase of the cell contents, that the most substantial movement of liquid takes place (as in a capillary tube). The movement may eventually develop into a mass flow. This explains why the pressure, the final balance between inflow and outflow, must prove to be lower for a membrane with wide pores than for one with narrower pores, and reaches a maximum when the pores become so narrow that salt no longer diosmoses. We shall leave until later a discussion of the relationship between membrane properties and the pressure.

As regards diosmosis through the mass particles themselves, essentially the same is true here as for the substances that are exchanged in the boundary layer under the influence of molecular forces, where there may also be cases when a substance does not osmose.

When we examine all the data available on the endosmotic equivalent (water/salt), there are none that cannot be reconciled with the principles developed here, once we acknowledge the role of the membrane properties. Since it was not my intention to do special studies on the endosmotic equivalent, however, I have no reason to discuss this topic in detail, and have to limit myself to a few basic points.

Relative exchange through porcelain cells or through other diaphragms that do not swell must always be simplest. Here too, obviously, a specific boundary layer will form, which --and this is important --must necessarily change to a certain degree as the concentration of the salt solution changes. There are other factors involved, but that one factor in itself is enough to explain why the endosmotic equivalent for different concentrations is not completely constant, and also why the salt flow may first increase more rapidly when the solution has low specific gravity, as Fick (1857:341) found,[41] and may even reach a maximum. In porcelain cells, filtration is relatively extensive, which is why a small pressure difference can cause a sizable error. Since in animal membrane and other membranes, including precipitation membranes, any pressure difference brings about filtration, this always causes a small error. However, the error is infinitesimal when the osmotic inflow is very large compared to filtration.[42]

In animal membrane and similar membranes, capillary pores as well as narrower spaces doubtless exist. The latter probably do not allow certain salts to diosmose. Such an arrangement could in itself explain all the observed phenomena of diosmosis and swelling, but these phenomena are also consistent with the simultaneous passage of water, or of salt molecules, through the membrane particles.[43] I must leave open the question whether the latter process is involved (as seems probable). As for the exchange ratio, note that certain spaces (or membrane particles), perhaps, only permit water movement that is produced in a unilaterally predominant way in the boundary layer, or at least in certain zones of the boundary layer. However, in animal membrane, not only is the boundary layer variable in the same way as it varies in procelain cells, but the diosmotic paths themselves must change their dimensions as the concentration of the salt solution changes (as their capacity to swell indicates). When, with changing conditions, the transition from one stationary state to another takes place over unequal time periods, there is no justification for using this argument as a basis for an inherent difference in the diosmosis between substances that can or cannot swell up as Fick (1857:296) tried to do. Incidentally, since nitrocellulose is not completely stable, there might be other factors involved with collodion membranes that keep a stationary state from being established, and perhaps delay it indefinitely.

I must add a short note on the osmotic behavior of certain acids, since many osmotic studies have not mentioned observations recorded by Dutrochet (1837:35) that were confirmed in principle by Graham (1854:225) and expanded by him, in part. I am speaking about the decrease in volumes of many acids when they are separated from water by certain membranes --the negative osmosis of Graham. Consider that the boundary layer and possibly the membrane particles themselves, depending on their mutual affinities, will be able to absorb solutes in relatively larger quantity than are present in the adjoining solution. Yet this process may well be reversed when the solution attains a certain concentration, thus the behavior of the acids alluded to is no longer surprising. It will also become clear how (for reasons given here and other reasons) that concentration, which produces no volume change at all, shifts. Thus different membranes can show quite dissimilar behavior (a fact already observed by Dutrochet for animal membrane as opposed to plant cell walls). In passing, let me note that solutions of oxalic acid and tartaric acid, with a concentration of 1% and 0.5% respectively, which according to Dutrochet would greatly decrease in volume when animal membrane is used, provoked a rather considerable water flow through a copper ferrocyanide membrane into the acid solution.

Until now, it has always been assumed that in the capillary spaces, which are not under the influence of the molecular forces extending from the membrane, salt and water are exchanged according to equal volumes. This is self-evident where there is diffusion in somewhat wider vessels. However, this is not necessarily so in narrow pores that connect two liquids with each other. Assume the paths travelled by two molecules that attract each other are in inverse ratio to the mass of the molecules (as in general mass attraction). Then, as soon as the density of the salt and water molecules is different, the volume will change on each side of a plane, thus producing a unilateral flow of liquid in the pore. True, general mass attraction cannot be regarded as the only decisive factor for actual molecular movement --the kinetic energy of the molecule itself,[44] and in our case other special circumstances also play a role here. But I did want to point out that a unilateral flow of liquid is possible (though it is not extensive) through capillary spaces that take up the salt solution unchanged.

11. Dependence of the Osmotic Water Flow on Membrane Properties and Solution Concentration.

Before we discuss quantitative experiments, it is a good idea to mention a few more conditions relating to precipitation membranes. Inevitably in our cells part of the total surface of the precipitation membrane has to be pressed against impermeable porcelain mass. Thus only the parts of the membrane stretched over pores of the porcelain cell are in contact with liquid on both sides. Therefore the parts of the membrane pressed to the impermeable porcelain have only a minor significance for diosmosis --i.e. insofar as liquid in the membrane moves parallel to the surface. As regards diosmotic action, the membrane area in contact with the porcelain mass can be equated only with a membrane of smaller surface that is free on all sides. This, incidentally, in no way detracts from the usefulness of our cells.

Diosmosing solutes need to travel not only through the precipitation membrane, but also through the porcelain cells. Since the cells are so highly permeable to water and salt, compared to the precipitation membranes, the influence of the porcelain cell can scarcely make itself felt. Thus, through a $100cm^2$ piece of copper ferrocyanide membrane under a pressure of 100cm mercury in the course of an hour, 0.04cc water were filtered in one case observed. While for an area of the porcelain cell of the same size, under the same conditions, the amount of filtration was found to be from 950-1300cc. Yet even if the smallest pores of the unevenly porous porcelain cell were actually to exert an inhibiting influence, this would not affect comparative experiments (at least when the effect of non-diosmosing substances is involved). Because, in narrow pores and in the precipitation membrane itself, the resistances increase porportionally to the amounts filtered.

In comparison to animal membrane, the amount of filtered liquid yielded by a disproportionally thinner copper ferrocyanide membrane is always only minimal. Thus, from experiments done by Schmidt (1856:348),[45] taking as a basis one of the lowest values and also the values assumed above, we can calculate the amount filtered for pig bladder as 8.87 cc. For other samples, it was found to be more than 200 times higher. In contrast to this, the filtration amount observed with our cells under the same conditions, which did not reach 0.04 cc, was minimal. It will remain so even if we use, as a basis for comparison, a membrane of the same area (free on all sides) instead of the surface membrane. True, in a membrane that is free on all sides, filtration appears to take place much more extensively. This can be inferred from the collapse of a copper ferrocyanide cell, closed on all sides, when it is placed into a sucrose solution of known osmotic strength.

The osmotic water flow through precipitation membranes is obviously proportional to the surface area of the membrane, but it is also proportional to the driving force. Since, as filtration experiments showed, resistances increase in the same ratio as the intensities of the flow. There is probably no question that the resistance opposing a water flow is proportional to membrane thickness.

As we already know, the membrane thickens when membrane-formers are present. This is why the water movement, corresponding to an osmotic driving force or a corresponding pressure, decreases. Also, it does this because of the plugging by foreign substances. To measure this thickening, approximately, by the reduction in the water movement, a cell with a copper ferrocyanide surface membrane was kept free from dust for five weeks, while filled with very carefully filtered 3% solutions of the membrane-formers. Immediately after forming the cell and at the end of five weeks, the water flow produced by a 4% sucrose solution was measured, at the same temperature (14.5°C). The inflow in the first experiment corresponded to a rise of 9.6mm per hour in the measuring tube (149.6 mm^3), and had fallen to 4.9mm after five weeks. Since there could only be an insignificant amount of plugging in this experiment, we may assume that in the intervening period the membrane had become at least twice as thick.

Earlier considerations showed how the osmotic driving force, developed in the diffusion zone, comes back to the same forces that cause the intermingling of salt and liquid in hydrodiffusion (p. 51). The intensity of this diffusion flow would, accordingly, permit us to draw a conclusion about the osmotic force of a substance, if the level and, more generally, the nature of the diffusion zone were known. Since Fick's studies it has been generally assumed that when the diffusion constant (k) -- i.e., the amount of salt that passes through unit area within unit time when the concentration decreases by one per unit length -- is known,[46] then the quantity of diffusing salt (s) is expressed (Mousson, 1871-74:282) by $s = k \cdot a \cdot t(e/l)$. Here (e/l) is the decrease in concentration over distance, (a) is the surface area and (t) represents time. Obviously, the same formula could be used to determine the water flow streaming in the opposite direction if k now represents the diffusion coefficient of the water. This assumed proportionality between concentration difference (e/l) and diffusion flow can, however, be correct or approximately correct only within certain limits, as the following reflection will show.

An increase of the diffusion flow proportional to the increasing number of molecules is possible only when all molecules that have newly been added exert equal forces of attraction upon adjoining water molecules. However, this will not be the case when, with increasing concentration, a certain number of molecules can no longer equalize their force of attraction to the water to the same degree. Thus, they will exist, with respect to water, as an unsaturated aggregation of molecules in solution. This aggregation obviously exerts a greater force of attraction on water molecules that come into its range. This is indeed so. In addition to other arguments, it is proved by the contraction which generally occurs when salt solutions and water are mixed, but which does not change proportionately with the number of dissolved salt molecules,[47] just as the density increases when salts are dissolved directly. However, as soon as the number of unsaturated salt molecules increases faster than the concentration, expressed in volume percent, the diffusion flow and the driving forces themselves must increase faster than the proportionality assumed up to that point demands, provided that other causes (such as the viscosity of the liquid, etc.) do not compensate for it.

Of course, for dilute solutions there will normally be only approximate proportionality between the concentration and diffusion velocity (see above). This, as well as the experimental methods, which do not permit a high degree of accuracy, may be the reason why a deviation was, until now, not definitely established.[48] In osmosis, however, this deviation may become especially large, because the difference in concentration in elementary layers that adjoin each other may be very great and it is possible that a concentrated solution may directly adjoin pure water.

Presupposing the same constitution of molecules and tagmas in solutions, the driving force behind the diffusion will grow proportionally to the number of molecules per unit volume. Correspondingly, the concentration must be expressed with regard to diffusion, by the amount of salt dissolved per unit volume, i.e., in volume percent. This was done by Voit, for example. Fick and Beilstein, on the other hand, expressed the concentration in percent by weight. Jolly expressed it as the quotient of dissolved salt divided by the water dissolving it.

Now let us take a closer look at the osmotic water flow which arises for different concentrations. To begin with, Tables 1 and 2, following, give an overview of results obtained with sucrose and gum arabic in copper ferrocyanide membranes derived from experiments (see section 17, Tables I and II). Column (c) shows the concentration of solution used in percent by weight, (e) represents the relative osmotic flow related to the flow of the one percent solution as a reference value. The column marked (e/c) contains the quotients of values for water flow listed in column (e) divided by the corresponding percentages by weight. Column e/(c•s) came about by dividing these quotients by the specific gravity (s) of the solution belonging to each. Since the product of the specific gravity and the percent by weight (c•s) gives percent by volume, e/(c•s) represents the quotient of the appropriate relative osmotic flow listed in column (e) divided by percent by volume. For one percent solutions, the quotient in this case remains e/(c•s) = 1/1.004, equals 1, when rounded off to the second decimal.

[The need for clarity and the constraints of format have led us to rearrange parts of the data presentation, for some tables, from that of Pfeffer. A list of symbols is provided here for the reader's convenience and to reduce repetition in the headings. Note that P replaces Pfeffer's O. The units for the data are as given here, unless stated otherwise. The data in Tables 1-14 come from Section 17, experimental results, Tables I -XIX.]

c = concentration in weight percent
e = relative osmotic flow (mean values)
f = filtration flow in mm^3/hr
h = rise of fluid in the manometer tube in mm/hr
p = filtration pressure in cm mercury
P = osmotic pressure in cm mercury
s = solution specific gravity

Table 1 - sucrose

c	e	e/c	s (17.5°C)	$\frac{e/c}{s}$
1	1.0	1.0	1.004	1.0
2	1.95	0.98	1.008	0.97
6	5.77	0.96	1.024	0.94
10	11.6	1.16	1.0404	1.11
16	20.0	1.25	1.0657	1.17
20	25.5	1.27	1.0832	1.17
32	48.4	1.54	1.1391	1.35

Table 2 - gum arabic

1	1.0	1.0	1.004	1.0
6	3.6	0.6	1.024	0.58
18	16.4	0.91	-	0.84

As the quotients of columns e/c and e/(c•s) show, the water flow is proportional neither to the amounts of salt contained in the unit of weight, nor in the unit of volume. With sucrose, the water flow increases, up to 6%, more slowly than the concentration in percent weight or percent volume, and continues to rise faster than the concentration. At 10%, there is already no doubt about this increase; however, it is very probably not uniform; only in general will the water flow increase faster than the concentration. The experiments we are looking at do not permit further conclusions: obviously, no significance can be attached to the quotient for 16% and 20%, which is the same (with respect to percent by volume), for this quotient was obtained from one experiment each.

There is a small decrease of the quotients e/c and e/(c•s) (true, it is only a slight decrease for a 6% sucrose solution); this decrease cannot be attributed to a mistake in observation, since the results of different experiments always pointed in the same direction. This decrease, which is only slight for sucrose, is highly conspicuous in gum arabic. One percent and 6% solutions of gum arabic show that the water flow initially increases far more slowly than the concentration, but then -- as the result obtained with 18% shows -- increases again with higher concentration, as also happens with sucrose. Yet even at this high concentration, the water flow never quite becomes intense enough to reach 18 times the output of a 1% solution.

The behavior I have just described cannot in itself be explained by the assumption that the resistance opposing (the water flow) in the membrane increases in a different ratio than the velocity of the water flow.[49] Moreover, direct experiments, which I shall refer to later, showed there was proportionality between the pressure and the amount of outflow. Furthermore, because gum arabic and sucrose do not diosmose, or practically do not diosmose, they only come into contact with the membrane surface, and therefore can in no case act on the nature of the membrane in such a way that the filtration resistance in the membrane is changed. The gratifyingly identical ratio between pressure and water flow for the same concentration of a substance, over several experiments, also speaks against such an assumption. This result, and the considerations I expressed earlier, also do not admit of a possible extension of the diffusion zone by the water flow. Also the above results themselves cannot be deduced from such an assumption.

That is why in each case the water flow that has actually been observed must be explained only by the osmotic driving force alone. This driving force can increase in a different ratio than the concentration, because it depends first of all on the nature of the diffusion zone and the force of the dissolved molecules that attract water. The latter must, as I explained before, increase more rapidly than the concentration if the number of salt particles in the solution, which could only imperfectly equalize their affinity to the water, rises more rapidly with higher concentration. For osmotic action, this circumstance can be of great significance. The nature of the diffusion zone will change along with the concentration of a solution. It is highly improbable, if not impossible, that the extension of this zone (resulting from the molecular forces active between water, salt, and membrane, and from the kinetic energy of the molecules) should always change to the extent needed for the production of a water flow proportional to the concentration.

In very dilute solutions, molecules of an additional quantity of salt probably combine with water essentially, though not exactly, in the same way as the molecules already present had done. The fact that the water flow increases more slowly at the beginning, and does not keep pace with the concentration, might be explained by the specific nature of the diffusion zone. On the other hand, when the osmotic water flow later increases, the accumulation of molecules (or tagmas) imperfectly saturated with water occurs more rapidly than the concentration increases and may be an important contributing factor. We must leave open the question of how these factors are involved separately and in combination, and finally how additional circumstances may also play a role here. Incidentally, I would of course not find it surprising if for certain substances a maximum osmotic water flow was observed even at a low concentration.

If a salt diosmoses, some special circumstances should be noted in addition to those already mentioned. These are involved in diosmosis and partially follow directly from our previous discussion (p. 55). Table 3 gives an overview of results obtained with a copper ferrocyanide membrane and potassium nitrate, a salt that diosmoses to a rather considerable extent.

Table 3 - potassium nitrate
(from section 17, Table III)

c	e	e/c	s	e/c•s
1	1.0	1.0	1.006	0.99
2	1.79	0.89	1.013	0.88
4	3.41	0.85	1.025	0.83
8	6.46	0.81	1.051	0.77
18	11.69	0.66	1.123	0.59

A similar result was produced by an experiment with potassium sulfate, which diosmoses only minimally. In a comparison of 1% and 4% solutions, a quotient e/c for the latter was found to be 0.83 when the quotient for a 1% solution is set at 1 (see section 17, Table IV).

The quotients e/c and e/(c•s), which decrease steadily, show how the water flow increases more slowly than the concentration. In fact, as far as the experiments permit us to judge, this is always true for solutions whose concentration is between 1% and 18%. Here again, as was to be expected, it was not possible to discover a simple relation between concentration and water flow,[50] nor was such a ratio obtained in experiments with collodion membranes and other types of membranes (Fick, 1857:322).

12. Osmotic Water Flow through Mixed Solutions.

Marignac (1874),[51] in recent times, has done studies on the diffusion of mixed solutions of salts that do not decompose. According to these studies, the diffusion velocity of salts in mixed solutions is not perceptibly affected. Graham had already reached this general result. He found (Graham, 1851:75 and 1862:30) that even the diffusion of crystalloids through viscous colloids was only somewhat slowed down. Therefore it would seem likely that the osmotic action of mixed salt solutions and the sum of the individual actions of salts do not differ greatly.

I had not set myself the task of solving the question I have just mentioned. Therefore I did only one experiment each with two different copper ferrocyanide membranes, using the same material, i.e., with gum arabic and with potassium nitrate, and with a mixture of both. The osmotic effect of these substances is almost the same in the mixed solution as in the unmixed solution, as Table 4 shows:

Table 4

solute	c	h (17.1°C)	h (15.8°C)
KNO_3	1	6.08	5.4
gum arabic	15	2.06	1.8
KNO_3 + gum arabic	1 and 15	7.9	7.0
KNO_3	1	6.06	5.3

(The gum arabic solution in this case was made from just air-dried material. The solutions contained no membrane-formers.)

The data of Baranetzky (1872:234) do not agree with the above results. According to Baranetzky's data, the addition of small amounts of colloids or crystalloids to, respectively, solutions of crystalloid or colloid substances is supposed to cause the water flow to increase considerably. True, these data have reference to other membranes (parchment paper, cellulose, etc.). Baranetzky even gives the osmotic effectiveness of the mixture (as regards unilateral increase in volume) as three times higher than the sum of the individual effects of the solutes. However, Baranetzky's assertion is not even true for parchment paper, and it becomes clear from the data, and because of other observations by Baranetzky, that his method must have at least been low in accuracy.[52]

For he claims that a solution containing 2% arabin and a solution containing 0.4% of certain crystalloid salts does not produce any osmoic increase in volume; he makes a similar claim for parchment paper with potassium nitrate. In contrast, I was able, with little effort, to observe a unilateral water flow for potassium nitrate solutions of 0.4% and 0.2%. Using an effective parchment paper area of 5.3 cm^2 and with 0.4% potassium nitrate solution in my measuring tube, the flow produced a rise of 7.2mm in the course of 8 hours. For a 2% solution of gum arabic, the rise was even greater, being 1.2mm per hour.[53]

It is true that Baranetzky worked with arabin, whose osmotic effect may indeed be lower than that of gum arabic (which still contains salts), but cannot be zero, because this solute practically does not diosmose, and any unilateral driving force and pressure causes water flow through parchment paper. Of course, with diosmosing solutes, even a decrease in volume is possible (see p. 57), but even for dilute potassium nitrate solution the experiment produces a definite increase in volume. Since I obtained the same result with two totally different types of parchment paper, it is not necessary to ascribe Baranetzky's different results to the nature of the material he used.

After the above criticism, we are forced to regard Baranetzky's measurements as inadequate wherever volume changes that are not quite large are involved. I see no need to go through all data in detail critically. Let me mention only one more point, i.e. (his statement) that a small addition of colloids to solutions of crystalloids (and vice versa) is supposed to cause osmotic output to rise significantly. In the following I did a comparative study of the effects of calcium chloride (1.5%) and gum arabic (2%) with parchment paper, and obtained the results listed in Table 5:

Table 5

Solute	c	h (1st)	h (2nd)
$CaCl_2$	1.5	9.9	10.3
gum arabic	2.0	1.2	1.3
$CaCl_2$ + gum arabic	1.5 2.0	11.4	11.3

(The experiments used different types of parchment paper; the area of the parchment was 5.3 cm^2 each time. For gum arabic, the observation time was extended to 5 hours. The temperature during the experiment was 17.4°C. The arrows indicate the sequence in which the experiments were done.)

The preceding figures show, as accurately as can be expected, equal effects for the components in the isolated state as well as in the mixture. Baranetzky (1872:239) also performed an experiment with the same concentration of the above substances, using parchment paper, only instead of gum arabic he used arabin, which was unavailable to me at the time. Baranetzky states that the volume increase over 24 hours was 0.5 cc for calcium chloride, and 0.9 cc for the mixed solution, while arabin was seen as producing no effect. The latter assumption, of course, is absolutely incorrect. In this case it is highly improbable that the use of arabin would lead to a result other than that I obtained with gum arabic. This, however, cannot be unreservedly decided offhand, because in certain cases, as we know, a type of chemical process can be imagined where the reaction products have a greater effect than their components. But if a chemical reaction or combination intervenes, then obviously the osmotic effects of the components and of the products resulting from these are not comparable, as we intend them to be. In any case, for simple mixtures, Baranetzky's assertion that the osmotic output of crystalloids is considerably increased by adding a small amount of colloids is definitely incorrect. Similarly, the converse assumption is incorrect that small amounts of crystalloids can increase the osmotic effect of colloidal solutes to a large degree. Two experiments which I performed in this direction for gum arabic and potassium nitrate turned out very much like the experiments listed in Table 5, which is why I do not think I need to communicate the details of them here.

13. Filtration under Pressure.

Table 6, which follows below, is based on experiments which were done with two different cells to determine the ratio between the pressure and the amount of filtration for copper ferrocyanide membranes. For the method used, see an earlier section (3), and for the documentation and the derivation of this table, see section 17. I shall merely state that the first column lists the level of the effective mercury column in cm, while the last column contains the normalized quotients of pressure divided by the amount of filtration.

Table 6

P (mean values)	f	f/p	f/p *(normalized)
210.2	18.5	.0880	1.023
208.0	17.5	.0841	0.978
112.2	10.3	.0927	0.992
111.5	10.5	.0942	1.009
85.1	7.56	.0888	1.033
71.3	6.54	.0917	0.982
37.8	3.52	.0934	1.000

*[*See Tables VI A and VI B for the calculation of these normalized f/p values.]*

As the experimental results section (17) shows in Table VI, these quotients reveal no larger differences, other than those that can be produced by the unavoidable errors. Also, the deviations fall evenly in both directions. I hope I may generalize beyond the limits of observation on the proportionality between pressure and filtration velocity demonstrated by these results. Specifically, I would like to point out that, as can be expected, there is no limit to the filtration resistance, i.e., any overpressure produces filtration in the direction of least resistance. Of course the amount of filtration produced by low pressure is very minimal. It is a good idea to use a pressure tube whose diameter is only fractions of a millimeter, so that the state of equilibrium will be produced within a short period.

As Poiseuille verified, proportionality between the amount of outflow and pressure is also true for narrow capillary tubes. When the width of these can be measured, the boundary layer is, however, always only an infinitesimally small fraction of the total diameter. Earlier considerations (p. 43) however, had indicated that in the precipitation membranes, even should the effective sphere of the tagmas not extend over the entire area of the intertagmatic spaces, no water molecule would still pass through without entering the effective range of the membrane particles. That is why, although the precipitation membranes are very thin, Toricelli's theorem plays no role here. This theorem is based on the free mobility of the liquid's molecules and on essentially the same principles as the free fall of a body in a non-resistant medium. On the other hand, as Poiseuille (1843) proved, capillary tubes below a certain length, which decreases rapidly with the radius, no longer satisfy the proportionality between pressure and outflow. A look at the following general equation set up by O.E. Meyer (1874) for outflow times (t) can explain the casual connection:

$$t = V \left\{ \frac{1}{\pi k R^2 \sqrt{2gh}} + \frac{8\eta\lambda}{\pi g h \rho R^2} \right\}$$

where V is the volume of the liquid, λ the length of the tube, R the radius, h the pressure, ρ the specific gravity of the liquid, η the coefficient of friction, k an empirical contraction factor, g and π have their usual significance. It is immediately clear that when the length is minimal the outflow, even for narrower tubes, approaches Toricelli's theorem whereas when the tubes are longer it approaches Poiseuille's law.

Willibald Schmidt (1856:367) observed approximate, but not completely exact, proportionality for pressure and the amount of filtration with animal membranes. I leave open the question whether the first factor in the above equation shows itself here because the capillary spaces are larger, or whether stretching of the membrane causes the capillary spaces to widen.

The well-known equation set up by Poiseuille for the outflow Q from capillary tubes

$$Q = \frac{C \cdot H \cdot D^4}{L}$$

(where H = pressure, D = diameter and L = length of the tube) cannot, as it stands, describe the filtration through a pore in the precipitation membrane. This is because C, for the same temperature and the same liquid, is not a constant factor but necessarily changes in an undeterminable manner with the molecular forces acting between the liquid and the membrane particles, and also with the width of the interstices.

The influence of the temperature on filtration through precipitation membranes will not be discussed until later.

14. Osmotic Pressure.

In a closed cell, according to our earlier definition, (p. 11), osmotic pressure is reached when in unit time equal quantities of fluid flow in and out. The osmotic driving force, which a non-diosmosing substance develops on the inner surface of the membrane, draws the water particles into the interior of the cell along the same paths as those on which they filter outward under pressure. The paths and the resistances are completely identical for inflow and outflow. Not so, however, if a solute is *[not]* capable of passing through the membrane. Here, the osmotic exchange in a pore can be quite insignificant for a predominantly unilateral water flow, if salt and water are exchanged in equal volumes, while pressure pushes fluid outward through this very capillary space (see p. 55). The state of equilibrium between inflow and outflow must then necessarily be established even at a lower pressure than in a membrane which does not allow the same substance to diosmose. If the latter is the case, the maximum osmotic pressure is reached. This osmotic pressure for a solute steadily decreases the more the diosmosis of this solute increases when there is an increase in size of the interstices in a membrane, formed of homogeneous material.

If there exists a more dilute solution in the variable boundary layer then, as earlier remarks point out, the water flow in this zone is not determined by the concentration of the porcelain cell content. Rather, it is determined by the osmotic driving force that has developed in the diffusion zone between the cell content and the boundary layer, subject to the difference in concentration.

For two solutes with the same membrane, one of them (provided it does not diosmose) will achieve its maximal osmotic pressure. While the maximal pressure will not be reached by the other solute that penetrates the membrane, and all the less the more it diosmoses. This is very forcibly shown by the following table, which gives the effect of 6% solutions with membranes of parchment paper, animal membrane, and copper ferrocyanide. Here, as in all following data, the pressure is always expressed in terms of the height of a mercury column measured in cm. (More details on the methodology of the experiments with parchment paper and animal membrane in section 17, Tables XVIII and XIX.)

Table 7

Solution	parchment paper	animal membrane	Cu_2FeCN_6
gum arabic	17.9	13.2	25.9
liquid gelatin glue	21.3	15.4	23.7
sucrose	29.0	14.5	287.7
KNO_3	20.4	8.9	?(700)

(The value given for potassium nitrate (700) is not directly determined; as far as I can judge by experiments with solutions with a different concentration, it is probably higher.)

Here, it is immediately evident that crystalloids (sucrose and potassium nitrate) at the same concentration, with a copper ferrocyanide membrane, produce a disproportionately higher osmotic pressure than colloids. However, the effect of colloids in animal membrane and parchment paper approaches the effect of the same substance with the copper ferrocyanide membrane, while with the copper ferrocyanide membrane the pressure for sucrose and potassium nitrate is 10 and 35 times as high. The reason for this is that the colloids, as we know, diosmose only very slightly through parchment paper and animal membrane, while crystalloids on the other hand diosmose very easily. The result is that in the experiment described above the following had diosmosed through parchment paper in the course of 3 hours, for the same membrane with an area of $5.3cm^2$ (these are approximate values): liquid gelatin glue (0.007g), a slightly higher amount of gum arabic, while at the same time about 0.14g sucrose and 0.55g potassium nitrate passed through the membrane. Incidentally, potassium nitrate also diosmoses noticeably through the copper ferrocyanide membrane and therefore does not achieve its maximal osmotic effect with a copper ferrocyanide membrane.

The necessary connection between osmotic pressure and diosmosis that I have set forth here is irrefutably shown by the above experiments. Yet, the values obtained do not give us a relative measure of the width of the pores in the different membranes, since the nature of the membrane material also plays a role in the osmotic effect. Still there can be hardly any doubt that the animal membrane used had wider pores than the parchment paper. The lower osmotic pressure, without exception, in animal membrane and the comparatively greater decrease of pressure for sucrose and potassium nitrate are thus explained without difficulties.

The low osmotic pressure that colloids produce with a copper ferrocyanide membrane is also shown by the following table (Table 8), in which the second column gives those pressures that were produced by the 1% solutions of solutes listed in the first column. Another experiment was carried out with conglutin.[54] The conglutin had been dissolved by means of as small a quantity of potash as possible. Here a pressure of 3.8cm was produced with a calcium phosphate membrane, while a 1% solution of sucrose with the same membrane produced 36.1cm pressure.

Table 8

solutions (1% by weight)	P (cm)	P_n (normalized)	f (normalized)
sucrose	47.1	1.0	1.0
gum arabic	6.5	0.138	0.14
dextrin	16.6	0.352	--
KNO_3	175.8	3.733	4.61
K_2SO_4	192.3	4.083	4.39

(The first data column has the mean values of measured pressures, the second has their ratio when the effect of sucrose is set at 1. The same ratio for the magnitude of the water inflow without pressure is given in the last column. The value for gum arabic has been derived from the series of experiments in section 17, Table VIII and is thus 47.1 times 0.138. The other experiments are described in section 17, Table IX.)

Dutrochet is the only person to have done comparative measurements on osmotic pressures in animal membrane with gelatin, sucrose, gum arabic and egg albumin (which also contains salt). Dutrochet's measurements do not indicate that colloids perform better,[55] but showed similar results for the commensurable substances, gum arabic and sucrose, as those listed in our Table 7 for parchment paper. Obviously, if we simply measure the intensity of the water flow, as Baranetzky did for colloids, we cannot simply jump to conclusions about osmotic pressure, and where osmotic pressure was determined with colloids, there are not comparative experiments with crystalloid solutes (Hofmeister, 1858:11). All these experiments were done with animal membrane, parchment paper or material with a similar behavior, and thus cannot give us an idea of the osmotic activity of the content of a plant cell, since this activity does not depend on the cell wall, as was assumed up till now, but is determined by the plasma membrane, which acts like a precipitation membrane.

A comparison of the relative performances, normalized to the effect of sucrose, with the water flow produced by the same solution, shows that for sucrose and gum arabic both pressure and water flow are in almost the same ratio. This, incidentally, was to be expected, since these substances do not diosmose. On the other hand, potassium sulfate diosmoses in smaller, potassium nitrate in larger, quantities through a copper ferrocyanide membrane. In both cases, but more so for potassium nitrate, certain paths are completely open to filtration toward the outside. Because these is salt solution inside, these paths are not fully available for the water flow that is directed toward the inside as they would be if the salt did not diosmose. This is why for the more easily diosmosing potassium nitrate the osmotic water inflow is relatively larger than for potassium sulfate, and still the final state of equilibrium for potassium nitrate is reached at a lower pressure than for potassium sulfate. The same ratio of pressure and water inflow in the non-diosmosing substances gum arabic and sucrose shows --as was already directly proved --that the amount of filtration increases in proportion to the pressure.

For the same reasons as with osmotic water flow, we cannot expect a simple ratio between the diffusion velocity of a substance and the pressure produced by the substance, the less so because in a possible diosmosis of a solute additional special circumstances are also involved. Only very generally will rapidly diffusing solutes also produce high pressures in precipitation membranes, i.e., the crystalloids produce more pressure than the colloids. The figures already given are quite sufficient to prove this; several less accurate experiments with other substances have easily confirmed it. Thus a solution containing approximately 0.3% anhydrous sodium sulfate, with a copper ferrocyanide membrane, produced the rather high pressure of 97cm, though the diffusion constant (Voit, 1867:233) value (0.527) is smaller than that of potassium sulfate (0.703). Similarly, in the same membrane the osmotic pressure for a solution containing not quite 0.3% ammonium acetate was found to be conspicuously high --87cm. At the same time, this salt diosmosed considerably, but it is also one of the solutes that diffuse rapidly.

Let us return once more to the relationship that exists between the width of the interstices in the membrane and the pressure, or between the pressure and the diosmosis of a solute. The osmotic driving force produced by a solute must (all other things being equal) remain the same as long as this solute does not diosmose, even if the size of the interstices of the membrane changes. Of course the amount of water flow increases as these interstices expand, but it increases in the same ratio for inflow and outflow.[56] These flows, after all, have to overcome the same resistances, so that the pressure throughout the process remains constant. The maximum pressure for a given substance is reached when the membrane interstices have returned to the particular width that no longer permits diosmosis to continue.

Unfortunately we have no way of varying the size of interstices in the same membrane at will. Aside from the minimal changes in the dimension of the interstices made possible by temperature fluctuations, other conditions that are significant for the osmotic driving force can always be changed. Be that as it may, it is interesting that the pressure produced by non-diosmosing substances, such as sucrose and gum arabic, with a copper ferrocyanide membrane, was raised only very slightly for a temperature increase of more than 20°C, as I shall describe later. The almost equal osmotic effect, which was produced by liquid gelatin glue with copper ferrocyanide and parchment membranes (Table 7), is explained by the fact that the interstices in parchment membrane (which, by the way, are relatively much larger), have approached the limit below which liquid gelatin glue no longer diosmoses. Only below this specific limit (if we imagine all other conditions to be constant) does the osmotic effect of a specific substance decrease as filtration velocity increases. But above this specific limit, pressure is constant, and will be able to give a measure for the molecular forces in effect between dissolved particles of salt and water, provided the other factors involved can be taken into consideration eventually. The strength of these molecular forces in crystalloid substances must be very great, as the high pressures produced by very dilute solutions indicate. In spite of this, the hydrodiffusion of these solutes is only moderately fast. The explanation for this is based on circumstances analogous to those which Clausius (1867:260,277) described, when referring to the diffusion and heat conduction of gases.

When solutes do not diosmose, the thickness of a membrane can affect only the velocity of the water flow, but not the magnitude of the pressure, as the above clearly shows. This would still be true even if (which is very unlikely) the water flow were not to change in proportion to membrane thickness. Thus the pressure gives a measure of the osmotic activity. This is independent of the thickness, surface area of the membrane, the number of effective molecular interstices per unit surface area and, finally, of the dimension of these interstices, as long as the osmotically active substance does not diosmose. If this is not the case, the pressure falls as membrane interstices continue to enlarge; but the pressure remains independent of the other quantities. It can be affected by the thickness of the membrane if, during capillary diffusion inside a pore, salt and water are not exchanged in equal volumes. Obviously, there is an assumption that the diosmotic properties of the membrane will not be altered, say, by the solute.

Here, I should point out that pressure can exert an influence on how much a solute diosmoses, although I did no such experiments. Let us assume that there are capillary spaces in a membrane in which salt and water are exchanged in equal volumes when there is no pressure difference. Then, as the osmotic pressure rises, a flow of liquid is driven outward, and therefore the amount of salt that leaves the cell per unit time must increase. The narrower the capillary spaces are and, therefore, the lower the velocity of the water flow becomes, the smaller will be the increase in the amount of salt leaving the cell, provided other conditions are the same. If inside the porcelain cell the salt diosmosed across the precipitation membrane is not removed immediately, the pressure would naturally have to fall, since there would be a dilute salt solution on the exterior surface of the precipitation membrane. Yet, I am sure this error was probably always extremely small, because of the only moderate diosmosis of the salts used and the high porosity of the porcelain cells.

The results obtained in the same cell with the same solution always corresponded to a high degree. Thus, with a copper ferrocyanide membrane for 1% sucrose solution, a pressure difference of 1cm was found only once, while in five other cells comparative experiments carried out with the same solution always produced an even smaller difference for one and the same cell. On the other hand, results obtained in different cells with an external copper ferrocyanide membrane differ from each other considerably. For a 1% sucrose solution, with which most of the experiments were done, pressures in 16 individual experiments measured between 47.1 and 53.8cm. This deviation cannot be caused by an error in method, but must be caused by the different qualities of the cell, for otherwise there was remarkably close agreement among experiments with the same cell that were often done very far apart in time. Yet the reason cannot be some sort of defective area, since it would have been repaired by the membrane-formers present. Also, there could not have been such complete agreement among experiments performed with the same cell.

I feel an obvious explanation seems to account for the deviation I have just mentioned. The term "external membrane" does not have to be interpreted in the strictest sense of the word; rather, the precipitation membrane, in fact, may very well be embedded inside the wall of the cell because of the way it was produced. That is, it may be covered wholly or partially by a thin layer of porcelain mass. But in the pores of the porcelain mass through which the sucrose solution reaches the precipitation membrane there is, in the boundary layer, a solution of a different composition, which in our case is probably more dilute. Where the precipitation membrane abuts on this boundary layer, the latter is, of course, active only by virtue of the concentration in this layer. Thus a lower pressure than one produced by a 1% sucrose solution would result. On the other hand, the resultant of all the individual outputs *[from the external and internal membrane areas]* is the actually measured pressure.

The pressure differences found for different cells are actually small enough to allow the explanation just given. To begin with, the differences will then depend on whether the precipitation membrane is completely, partially or not at all covered by the porcelain mass. The physical and chemical properties of the porcelain material will also be a factor as well. Once the precipitation membrane is embedded inside the porcelain mass, it is easy to see that it must make no difference for the pressure whether the porcelain mass that covers the membrane forms a very thin or a thick layer. The latter is the case for the precipitation membranes embedded in the middle of the porcelain mass. I performed only a single experiment with such a copper ferrocyanide cell. The experiment yielded a pressure of 47.0cm for 1% sucrose solution, i.e. the same pressure found to be the lowest value for external membranes.

Because of the unequal pressure, the results obtained with different cells are only comparable among themselves if they are reduced to a normalized value. When I compared the osmotic effect of different substances, I used the pressure resulting from a 1% sucrose solution as the basis of this normalization. True, such a comparison is not completely without errors, because the reduction of the osmotic effect, which was produced according to the above principle, is hardly proportional to the pressure produced by different types of solutions. But it is easy to see that the error cannot be great, and this fact also emerges from the ratio of the values obtained in different cells for the same solutions. There is no need for special explanations of how the conditions I have just described are involved in the osmotic water flow that takes place without pressure.

The only other precipitation membranes I used, in a small number of experiments, were some produced from prussiate of iron or calcium phosphate. The mean of two experiments, done with different cells, for the prussiate of iron membrane gave a pressure of 38.7cm for a 1% sucrose solution. A single experiment with calcium phosphate membrane, for the same solution, gave a pressure of 36.1cm (see - section 17, Tables XVI and XVII). Here, the diosmotic behavior of these two membranes agrees with that of the copper ferrocyanide membrane, insofar as demonstrable traces of sucrose also diosmose through those two membranes only when concentrated solution is used. That is why it will probably be permitted to ascribe the lower pressure to the fact that the composition of the diffusion zone is less favorable.

Now let us look at the osmotic pressures produced by solutions of the same substance at different concentrations. In the following tables, I have put together the results obtained with sucrose and gum arabic. The second column gives the pressures measured in two different cells, from which the mean values, after normalizing to the same value, were obtained and listed in column three. The column headed P_n/c lists the quotients thus characterized, while the last column, e/c, contains the corresponding quotients for the water inflow (without pressure) divided by the concentration (see Tables 1 and 2, p. 63). The experiments with sucrose, documented in Table VII of section 17, were carried out at temperatures ranging between 13.5° and 16.1°C. Experiments performed with gum arabic are documented in Table VIII of section 17. Where several values obtained with the same cell were available, for the same solution, the mean value was listed in Table 9.

Table 9 - sucrose

c	P	P_n	P_n/c	e/c
1	53.5 and 47.2	1.0	1.0	1.0
2	101.6	1.9	0.95	0.97
2.74	151.8	2.65	0.97	-
4	208.2	3.89	0.97	-
6	307.5 and 276.9	5.71	0.95	0.94

Table 10 - gum arabic

c	P	P_n	P_n/c	e/c
1	7.1 and 6.7	1.0	1.0	1.0
6	27.5 and 24.3	3.75	0.62	0.6
18	120.0 and 118.4	17.28	0.96	0.91

The quotients of pressure divided by concentration (P_n/c) and water inflow divided by concentration (e/c) are almost identical, especially for sucrose and nearly so for gum arabic. That is, pressure and water inflow increase in the same ratio. What I said earlier about the relation between concentration and water flow is thus also true of the relation between concentration and pressure.

In gum arabic the quotients (e/c) are somewhat smaller than P_n/c and from this it would follow that the water flow for solutions of different concentration, of this substance, increases somewhat more slowly than the pressure. The slight differences could very well be due to sources of error. We are even less justified in drawing definite conclusions because these values are derived from an inadequate number of observations. If these quotients were free of error, then they would demonstrate, for one thing, that the diffusion zone is extended by unilateral water flow. Also, by this extension of the diffusion zone, flow would be relatively slowed down. Of course, a considerable extension of the diffusion zone is not possible, as earlier considerations indicated. Yet if anything, we can expect a slight extension with gum arabic and similar substances, for here the viscosity of the liquid, and the fact that the force that works to keep the diffusion zone constant is perhaps relatively less strong, both play a role.

As for the diosmosing salts, I did only a few experiments with different concentrations of potassium nitrate, which did not turn out quite to my satisfaction. A few experiments performed with the same membrane and the same solution (which I shall not describe here) produced values that were considerably different from each other. Also, several times the pressure, once reached, again fell a great deal, obviously because the precipitation membrane became damaged, somehow, though I do not know why. That is why I would not like to attach any particular importance to the results listed in the following table.

In the table, the second column again shows the pressures measured, the third lists the quotients (pressures divided by concentration) and, finally, the last column lists the quotients calculated when the effect of a one percent solution is set at 1 (section 17, Table X).

Table 11 - KNO_3

c	P	P/c	P/c
0.80	130.4	163.0 }	1.0
0.86	147.5	171.5 }	
1.43	218.5	152.8	0.91
3.30	436.8	132.4	0.79

The quotients of water inflow divided by concentration were determined to be 1, 0.89 and 0.85 for 1%, 2% and 4% solutions of potassium nitrate (see p. 66). I shall not derive approximate values from the above results for those concentrations whose osmotic pressure was measured for, as I said, I do not consider these measurements to be very accurate. For that reason the quotients (which clearly do not exactly agree) would not permit us to reach any definite conclusion.

15. Fluctuations of the Pressure.

Up to this point, I have not discussed the influence of temperature since, in fact, as the following shows, our experiments or series of experiments -- where temperatures only varied by a few degrees -- could be compared with each other without perceptible error.

Provided chemical actions are not involved, a rise in temperature will increase the mean distance between the finer constituent parts in the membrane, but also between the molecules aggregated into a tagma, while at the same time it will reduce cohesion and viscosity in the liquid. Also we can generally expect that there will be less adhesion between the wall and the liquid. Such a decrease also appears when capillary rise falls with the temperature, and the variable boundary layer (as well as the diffusion zone) may also change in terms of its expansion and structure, depending on the controlling molecular forces. At present it is impossible to ascertain what these changes are or to derive them from the determining factors, since this is not even possible for the far simpler case of free diffusion, which is known to be very much accelerated by a rise in temperature (Graham, 1862;27 and Fick, 1866:28). In addition, in diosmosis the totality of the molecular forces acting between the membrane on the one hand and the solvent and solute on the other is an extra contributing factor. The process must be even more complicated if chemical rearrangements are involved as well.

Let us examine the simplest case -- a solute that does not diosmose at any temperature during the experiment. As the temperature rises, water flow is naturally increased for equal driving force. Now, because the resistances and paths for inflow and outflow are the same, the pressure only changes when the ratio between the forces that drive the two flows changes.[57] In our case, a rise in pressure will indicate an increase, while a lowering of pressure will indicate a decrease of osmotic driving force.

Sucrose and gum arabic are substances that do not perceptibly diosmose through the copper ferrocyanide membrane used, at the temperatures which are listed below. The exact proof of this was obtained by checking the specific gravity of the solution and by returning to the former pressure and reestablishing the original temperature (see section 17, Tables XI and XII). For both these solutes, pressure was determined at temperatures that were very far apart, and the results obtained were put together below. Three series of experiments (section 17, Table XI) were carried out with sucrose in three different cells. In each individual series of experiments, the cell remained unopened during temperature fluctuations. When pressure was determined with the same cell and solution, which did not vary greatly, only the mean value of the temperature and of the pressure is given in Table 12.

Table 12 - 1% sucrose

	temperature - °C	pressure (cm)
XI A	*14.2	51.0
	32.0	54.4
XI B	6.8	50.5
	13.7	52.5
	22.0	54.8
XI C	15.5	52.0
	36.0	56.7

*[*from XI A, e.g., (13.5 + 14.8)/2 = 14.2 average value*]

Table 13 - 14% gum arabic

XII	13.3	69.2
	36.7	72.4

As can be seen from these tables, the pressure rises slightly with temperature. Even if this rise is not large, it is still impressive enough to document this fact, since it reappears constantly in every series of experiments. Also, the experiments done with sucrose indicate that the pressure increases continually between 6.8° and 36°C. However, for differences of 2° and 3°C, this increase is too small to cause appreciable errors, even if neglected. I cannot say with certainty which particular variables cause the slight increase in pressure.[58]

The table that follows shows that as temperature rises the osmotic water flow becomes considerably larger. The table gives water flow, measured at different temperatures, with a 5% sucrose solution.[59] The second column lists the levels in the measuring tube in millimeters, observed over a one hour period, while the last column contains the quotients indicated by that heading.

Table 14 - 5% sucrose

temperature - °C	h (mm/hr)	h/temp.
7.1	5.9	0.831
17.6	9.4	0.534
32.5	13.3	0.409

(The first observation was made at 17.6°C and so was the last. For this temperature, inflow values were 9.5 and 9.3 per hour. The observation period was extended in such a way that at every temperature a rise of at least 12mm in the measuring tube was recorded. When there was a change to a different temperature, the cell was filled with a new solution each time.)

A simple ratio beween temperature and water flow cannot be derived from the above figures, and could not be expected right from the start. These experiments are not enough for us to derive an interpolation formula from them. Also, the limited usefulness of such an empirical formula seemed not to justify my carrying out a larger number of experiments.[60] Also, I refrained from determining whether solutions of other substances show the same water flow ratio, at the same temperatures as those listed in Table 14. Presumably small deviations would be found. However, where temperature differences are slight (as they were in the comparative water flow experiments described earlier), deviations are no doubt too small to be able to affect the result very much.

Let me also add that comparative experiments in darkness and in bright diffuse light showed no measurable difference either for the intensity of the water flow or for pressure. Incidentally, all my experiments are always done in dim light or with the light off.

A direct measurement of the filtration velocity at variable temperatures was not carried out accurately enough to be worth mentioning here. Since the pressure created by sucrose fluctuates only little with temperature changes, the inflow values listed above also give an approximate measurement for the filtration velocity ratio at different temperatures. Naturally this filtration velocity could also be exactly derived from measuring the osmotic water flow and the pressure.[61]

It is possible that the diosmosis of a particular substance does not begin until the mean distances between membrane particles become larger. Speaking very generally, however, the diosmosis of a solute will increase with temperature, particularly as a result of the increased active force of the molecules and the widening of the interstices within membranes. However, this speeding-up, like the water flow, will not be in a simple ratio to the rise in temperature. For obvious reasons, the composition and extension of the boundary layer will vary with temperature. If this is true, the diosmosis of a solute will not have to increase with temperature in the same ratio as the water flow going in the opposite direction, except for the fact that other variables can also be operative in the same sense. Eckhard (1866:67),[62] of course, found the endosmotic equivalent of sodium chloride to be constant over a wide temperature range, but in this case one salt is not representative of others, and the properties of the membrane must also be taken into consideration. Obviously, the smaller the area of the interstices that do not lie within the effective sphere of the membrane particles, the more obviously an eventual influence of the variable boundary layer can make itself felt.

Variations of the pressure are especially interesting from a physiologist's point of view. Understandably such variations can be produced by changes both in the membrane and in the cell content. For solutes that do not diosmose initially (as we said earlier) a simple widening of the membrane interstices only produces a variation in pressure if such widening leads to diosmosis. Also, an increase or decrease of pressure can be caused by such changes in the membrane, which influence the composition of the diffusion zone. As for the osmotic effect of the solution in the cell, it may vary not only through concentration changes, but also through chemical rearrangement.

The conditions for pressure changes, which I have described here, can obviously be brought about in a number of ways, particularly when a substance enters the membrane or the solution in the cell, or by means of reactions caused by forces from outside. As for the latter, physical and chemical data, especially where heat and light are concerned, offer us important clues that may lead to productive ideas in the field of physiology. I shall come back to this in the physiological section. The discussion of this that follows should be judged from that standpoint: it is not meant to be conclusive, but to offer new ideas and draw the reader's attention to selected theoretically important points.

Probably the simplest situation we can have is when a saturated solution, along with undissolved material of the same solute, is present in the cell. Depending on whether the solubility increases or decreases with temperature, pressure must then rise or fall as the temperature is raised. I only give an example for this self-evident variation -- which can be repeated any number of times -- because I did such an experiment. Excess potassium bitartrate was placed in a cell that had an external copper ferrocyanide membrane. After the pressure of the solution, which was saturated at 13.0°C, had been determined to be 68.3cm, the temperature was raised to 29.2° and the pressure was observed to rise to 115.8cm (see section 17, Table XIII). Incidentally, potassium bitartrate dissolves in 240 parts of cold and in 14 parts of boiling water.

Also of interest to us is the dissociation of liquid or dissolved substances which, for instance, only begins or is accelerated when light or heat increase the kinetic energy of molecular movement and disaggregation. The more complex the structure of a dissolved particle and the weaker the forces that hold it together, in general, the easier it will be for an increase in the kinetic energy to bring about a separation into simpler molecules. Thus, the more will the number of dissociated molecules increase (Naumann, 1876-77:488, 514 and Pfaundler, 1874). Of course for each temperature, for instance, a state of equilibrium sets in, and in this state during unit time as many particles are dissociated as are newly formed by the collision of molecules. While previously a particle -- such as a tagma -- could only be considered as a uniform whole, after dissociation it is the separated molecules that are active. Thus, when other properties of the solute change, it would be a coincidence if the osmotic effect remained the same. Normally it can be expected that where molecules separate from their compound by dissociation, pressure will increase because the large size of particles, tagmas or molecules is unfavorable to diffusion velocity and thus also for osmotic effect (all other things being equal).

There are two principles involved in dissociation processes, as such, brought about by temperature: either water is separated from the dissolved particles or the latter decompose, forming a base or a basic salt (Naumann, 1876-77:551). These processes could probably be best, and more generally, differentiated as tagmatic or molecular dissociation (a breaking up of the tagma or the molecule). The solubility maximum of sodium sulfate ($Na_2SO_4 + 10 \cdot H_2O$), a result of the formation of the anhydrous salt (Naumann, 1876-77:480 and Naumann, 1869:76), is one of the numerous dissociation phenomena of this type. Cobalt chloride strikingly demonstrates it, since at a higher temperature the red solution turns blue because the water-containing molecular aggregate dissociates. Similarly, molecular dissociation by precipitation of iron oxide is directly demonstrated when dilute iron chloride solution is heated. It is interesting that, for a reason that is obvious, the decomposition temperature in the latter case is lowered when the solution is dilute while, on the other hand, it is raised when the water of hydration separates.

The degree of dissociation is dependent on the temperature and on the specific properties of the solute. The extent of dissociation is also determined by the pressure to which the liquid is subjected. This dissociation must, in general, be slightly inhibited as this pressure rises, (in the case that the products of dissociation occupy a larger volume - which is normally the case) and, conversely, be accelerated when the volume decreases as dissociation takes place. This happens for the same reasons as those that underly the increase in the melting temperature of paraffin and, conversely, the decrease in the melting temperature of ice, with pressure. As the general equations of Clausius

(1876:172) show, the foregoing simply depends on whether the difference between the specific volume of the solid and of the liquid state of aggregation is positive or negative.

The few experiments have not given us satisfactory results as regards the direct importance of dissociation for pressure fluctuations. We should not forget, however, that it is important to choose an appropriate solute. I shall continue by suggesting how indirect releasing actions of the products of dissociation might be of first importance for physiological processes.

An experiment with a solution containing 0.3% anhydrous salt of sodium sulfate with a copper ferrocyanide membrane produced, at first, a pressure of 95.2cm at 14.3°C. When the temperature was raised to 35.5°C, the pressure rose to 98.9cm, and when the temperature went down to 14.8°C, the pressure was at 91.8cm, for salt had diosmosed during the six days it took to complete the experiment. That is one reason why sodium sulfate is an inappropriate choice; another reason is that because of its large osmotic effect it is necessary to use a dilute solution, in which the dissociation (i.e., here the splitting off of water of hydration) is suppressed. The above experiment does show one thing -- that a dilute solution of sodium sulfate, over temperatures ranging from 14° to 35° does not cause a large fluctuation of pressure.

The relatively strong diosmosis of ammonium acetate, which also produces a high pressure, caused an experiment with this salt to turn out unfavorably. Because of diosmosis, the initial pressure for a 0.3% solution in the cell was 87.5cm; six days later, at the same temperature (14.2°C), it was determined to be 67.0cm. On the other hand, the pressure found at 36°C was 85cm. It is impossible to say to what extent loss of salt counteracted an expected effect of dissociation. At any rate, the experiments do show that here, too, the increase of pressure with temperature cannot be very great. Dibbit claims that at 100°C a solution of ammonium acetate has 7% of the salt dissociated, but his claim is not based on reliable experiments.[63]

A few experiments were also carried out with double salts, though according to the thermochemical studies of Favre and Valson (Naumann, 1876-77:535), and the diffusion experiments of Marignac (1874),[64] they supposedly do not exist as such in solution.

Two experiments with sodium potassium tartrate (tartarus natronatus) gave the following result (section 17, Table XIV). Pressure, for a 1% solution at 13.3°C, was determined to be 147.6cm. At 36.6°C it was 156.4cm, while in the meantime the concentration had fallen by 0.06% because of diosmosis. The pressure, for a 0.6% solution at 12.4°C was 91.6cm. At 37.3°C it was 98.3cm and the final pressure, at 14.2°C, was 90.0cm. Thus, as temperature increased, the pressure for the 1% solution had risen by at least 8cm, while it rose at least 7cm for the 0.6% solution.

If we compare this increase with the smaller increase in pressure found for sucrose and gum arabic when temperature increased, the idea suggests itself that dissociation may be involved here; however, these results are not at all definitive. They could perhaps become so if the osmotic pressures for potassium tartrate and sodium tartrate could be determined individually, at the corresponding temperatures.

The same is true for sucrose - sodium chloride ($C_{12}H_{22}O_{11}$ + NaCl). A solution containing 1.171% of this double salt gave the following pressures: at 14.5°C, 123.6cm; at 37.9°C, 129.2cm, and at 15.0°C, 120.7cm (section 17, Table XV). The falling pressure, while the experiment was going on, was produced by the diosmosis of a small quantity of sodium chloride.

An experiment carried out with a prussiate of iron membrane, with iron alum, gave a result that was not sufficiently satisfactory -- specifically a rather considerable amount of iron oxide had been precipitated in the course of several days. Perhaps the cause for this precipitate is a diosmotic transition of the dissociated hydrochloric acid into the surrounding liquid, and the progressive dissociation of the iron salt this causes.

The experiments with sodium potassium tartrate and sucrose-sodium chloride described here cannot directly ascertain whether these double salts partially exist in aqueous solution. This seems likely, also, for another reason, i.e., the state of motion of matter. Because the state of motion is variable, the number of existing molecules or tagmas of the double salt probably also changes. As the kinetic energy becomes greater, and thus as temperature rises, this number must probably decrease. The experiments carried out until now are, incidentally, not accurate enough for us to be able to find out with certainty whether a small amount of the double salt continues to exist in solution. If we follow the comparative procedures indicated above, the measurement of osmotic effectiveness in precipitation membranes provides a new method for discovering how much dissociation there is without having to separate the dissociation products from each other.

Fluctuations in pressure when chemical transformations take place inside the cell are to be expected, because different substances have different osmotic effect. So if the above experiments did not produce the desired result, the reason is that suitable decompositions were not produced.

Processes of dissociation by heat and especially by light are numerous, and many examples are known. Little research has been done specifically with solutes whose particles have complex structures and play an important role in the organism. For example, the disproportionately higher osmotic effect of crystalloids shows how much the pressure must fall if, as a result of increased molecular movement produced by light or heat, the tagmas of a colloidal substance were split into crystalloid molecules. In just the same way, through this type of process or similar ones, the diosmosis of a substance could be modified or actually even initiated.

Only a small fraction of a substance will be dissociated by heat or light. The steady-state condition, in which the same number of particles dissociate as recombine, is soon reached. But if one of the products of the dissociation is continuously removed, whether through chemical binding, diosmosis or in another way, there is a possibility that, also, a minimal rate of dissociation may lead to total decomposition. This point of view has many possibilities with regard to osmosis -- both diosmosis and pressure. I would like to speak about only one possible case, since anyone who is familiar with the pertinent physical and chemical facts can easily formulate a large number of questions, some of which are physiologically important. When iron chloride is exposed to light, only a small amount of it dissociates. If the hydrochloric acid that has become free were continuously removed by diosmosis with a suitable precipitation membrane, the dissociation would proceed. According to what Graham (1862:45) tells us about the formation of soluble iron oxide hydrate,[65] we can predict with certainty that a colloidal ferric hydroxide, containing only a little hydrochloric acid, would be the end product. However, by adding hydrochloric acid, i.e., by placing the cell (which remains closed) in water containing sufficient hydrochloric acid, the initial condition could once more be reestablished. But as colloidal iron oxide forms, the pressure would probably decrease a great deal, since iron chloride has a very high osmotic effect, at least with a prussiate of iron membrane. In general, as we know, the effect of light and heat first disrupts the association of the molecules. As a result of this, powerful reactions can come about when other substances are present. We all know how easily the combination of chlorine and hydrogen takes place in the presence of light, accompanied by an explosion. This combination does not take place until a number of chlorine molecules are dissociated into their atoms, as directly proved by Budde (1872 and 1873). Similarly, as a result of the loosening of the molecular association, in light, higher oxidation states of metal salts pass into lower oxidation states when oxidizable substances are present. Related to such a simultaneous oxidation-reduction process is the development of CO_2, sometimes very active, when a mixture of iron chloride and oxalic acid is exposed to light (and not until then), see Becquerel (1868:71). Obviously, osmotic processes can be modified by such decompositions in a large number of ways and, for instance, when reducing mercuric chloride, a lower pressure would have to be expected as the insoluble mercurous chloride was precipitated.

Especially important for physiological problems are those chemical transformations which are brought about by a comparatively small amount of an active substance, i.e., if need be, by a minimal quantity of a substance set free by dissociation. Not until Williamson explained how ether was formed did people begin to understand the mysterious, so-called, "catalytic" effects. The constant reaction between ether sulfuric acid and alcohol, in ether and sulfuric acid, and the constant new formation of ether sulfuric acid from alcohol and sulfuric acid, this continuity of two processes taking place side by side, enables us to convert a large amount of alcohol into ether with a little sulfuric acid. Indeed, the amount of ether that could be produced with a minimal amount of acid would be unlimited if the reaction went to completion. We must regard the behavior of the so-called ferments, which can chemically transform a disproportionately large amount of a substance, in an analogous fashion, i.e., as the result of the continuity of chemical processes. It is precisely these ferments that produce wonderful chemical processes in the organism. It makes no difference whether such ferments penetrate the cell (or another organ) from the outside, whether they are freed and activated by dissociation, or whether the ferments themselves are already present but need an additional substance to activate them -- whatever the case may be, it is clear that even, say, minimal dissociation can cause very far-reaching chemical changes, and thus can give new directions to the osmotic processes and effects as well.

For the osmotic processes, however, not only the content but also the properties of the membrane are important. Light and heat, for example, can produce certain effects without chemical intervention; but the latter can also become significant unless some agent influences the membrane directly, or by means of decomposition products produced in some other way. For example, the gradual decomposition of prussiate of iron when exposed to light studied by Chevreul (see Becquerel, 1868:72), and the extraction of phosphoric acid from iron phosphate by means of alkalis are processes that could be carried out without destroying the continuity of the precipitation membrane. Moreover, as Traube showed (1867:141), the osmotic property of a membrane can be changed by infiltration, i.e. by the embedding of foreign particles in the membrane. Infiltration with barium sulfate, for example, is said to make a membrane of tannic gelatin glue impermeable to ammonium sulfate.

Primarily, however, the fluctuation of osmotic processes brought about by physical or chemical change in the cell content or the membrane is only of minor physical interest,[66] but of all the more physiological interest for us. The physiological process itself will first have to provide us with questions for our experimental studies, studies by means of which we can then perhaps discover the truth about processes that take place in the organism. I am sure the osmotic processes we see in the organism are always the result of various complicated separate effects, and that is precisely why it is tremendously difficult to reduce the visible effects to their causes and to explain them once these causes are known. Because the processes are so complex, physiology is a very broad field promising great success to the researcher who has the ability to overcome the difficulties of the subject.

16. Historical Review.

Since the relevant literature essentially was cited at the appropriate places in this paper, I shall simply give a short historical review here.[67]

Osmosis was discovered as early as 1748 by Nollet, but so little attention was paid to the discovery that the later rediscovery by Fischer in 1812 seemed like a new discovery.[68] The data presented by those scientists mentioned above and by Parrot, whose work was based on Nollet's discovery, were not the result of investigations that strive for deeper insight or are dominated by guiding concepts. It was Dutrochet who published numerous studies on osmosis between 1826 and 1837 and who also attempted to give almost as many different explanations for the phenomenon, of whose great significance for physiology our author was very much aware. Since Dutrochet himself later rejected, especially, his attempts to explain osmosis on the basis of electricity and capillarity, and no other innovative work appeared between Dutrochet's first publication and his last summarizing paper,[69] I feel justified in quoting only his final corrected views here.

Dutrochet's endosmometer consisted of a glass tube simply closed on one side with a membrane (this apparatus had basically already been used by Nollet and Fischer). Subsequent apparatus used for osmotic experiments is also based on this principle. Dutrochet essentially only measured the volume change of the cell content, and the pressure produced by given solutions, but did not quantitatively determine the amount of solute passing into the outside liquid. Dutrochet did recognize that the properties of the membrane are decisive for the diosmotic flow of both water and of the solute, and that furthermore this exchange depends on the concentration and the temperature.

Let us first leave aside theoretical considerations, and look at experimental work after Dutrochet. The studies of Jerichau, Kürschner,and Mateucci and Cima did produce a great variety of data, but basically discovered no new principles or methods. Vierordt (1848) then measured the exchange of solute and of water quantitatively, with an apparatus that probably permitted greater accuracy than the endosmometers used in the period immediately following his work. But while Vierordt emphasized the main importance of measuring, per se, and especially of measuring the water flow, Jolly (1849) stressed the value of determining the ratio between the amounts exchanged. He, as we know, called the quotient of salt divided into the amount of water, simultaneously passing inward, the endosmotic equivalent. To determine this endomotic equivalent was thereafter almost the exclusive purpose of numerous studies. Jolly's view that the endosmotic equivalent was a measure independent of such things as concentration was, it is true, very soon disproved by Ludwig (1849). Other studies (e.g., those by Fick, Schmidt (1857) and Eckhard),[70] also threw light on this and other aspects. This is not the place to discuss in detail the endosmotic equivalent and questions relating to it, since, as I said earlier, this topic was not considered in this work. The experiments studying the osmotic water flow, in particular, have been mentioned here wherever necessary. I have also already stated that there are no later studies on pressure that elaborate on the results already obtained by Dutrochet.

Graham (1854), whose first work on osmosis brought little that was really new and insightful, then paved the way for something very innovative by making the distinction between crystalloid and colloidal solutes, and establishing that the latter diosmose very slightly (Graham, 1862). One wonders why, on the basis of this, no one considered the particular case where the osmotically active solute does not diosmose, and looked at it as the starting point of the theory. That was actually not done until Traube (1867 and 1875), whose discovery of precipitation membranes is one of the most important, if not altogether the most important, steps forward since the discovery of osmosis.

Dutrochet's views on osmosis and its cause, which took their final form after many modifications, do not offer a deeper analysis of the phenomenon, it is true, but do express general principles whose validity is still recognized today. Dutrochet (1837:58) requires that these prerequisites are imperative in order for diosmosis to be possible: 1) that at least one of the two separated liquids must have an affinity for the membrane, and 2) that between the two liquids there must be affinities which can produce a mixture. In addition, and mainly because of the opposite osmotic behavior of alcohol and water with animal membrane and India rubber, Dutrochet stresses that the solute that passes through a membrane in greatest quantity is the one that has the greatest affinity for the substance of the membrane. He also says that the exchanging substances are not separated in the membrane, but contained in it in mixed form.

Dutrochet's osmotic studies have been condemned, more than once, groundlessly I feel, if we look only at the bare facts or at Dutrochet's views which, finally, were so clear. His fluctuating between one theory and another is not isolated to this case but is closely related to his intellectual character. A skilled experimenter and good observer, Dutrochet often found his wealth of ides to be an obstacle to finding a realistic viewpoint for the world of phenomena revealed in his osmotic studies and other research. Enchanted by an idea which seemed to place the final goal of his research and the general concepts he strove for within his reach, Dutrochet, who could at other times be so perceptive, often overlooked questions whose critical consideration would have put a stop to his flight of ideas. Frequently, after initially forcing facts that he had perceived correctly into an unnatural framework, he later allowed those facts to lead him along the right path. Under such circumstances he actually deserves to be harshly criticized for particular points; however, if we judge him not for these details but for his overall achievements, Dutrochet must be celebrated as a scholar of great importance, especially for his work in plant physiology.

After the views of Poisson and Magnus (see Vierordt, 1846:507), based on capillarity, had very soon proved to be inadequate, Brücke (1843, also in his 1842 dissertation) was the first to establish a more detailed theory that still adequately accounts for actual cases. As we learned earlier, this theory postulates a boundary layer which varies with the distance from the pore wall. Brücke demonstrated how it was formed, with mixtures of turpentine oil and olive oil, using a suitable apparatus. This theory is still true without reservations for simple porous material, which does not swell. But, we must add that the boundary layer varies with concentration and is not constant, as Brücke and others after him seem to assume. On the other hand, Brücke neglected the change in the diameter of the pore, unavoidable because of swelling. Especially, he did not take into account at all diosmosis through the membrane particles themselves (diatagmatic diosmosis).

Ludwig (1849) then demonstrated that animal bladder imbibes a more dilute solution from a salt solution.[71] This, it is true, seems to indicate a boundary layer consisting of dilute solution (in the sense of Brücke), but is not a completely convincing argument, as Ludwig and others after him assumed, since the absorption of water or dilute solution into the membrane particles themselves would have the same result (p. 40). Ludwig already pointed out some inferences based on Brücke's theory and confirmed by experimentation the theoretically postulated change of the endosmotic equivalent as concentration changes.

Later, Fick (1855:74) deduced in detail the consequences of Brücke's theory, and even though the contradictions he found between theoretical reasoning and experimental findings are basically founded on an interpretation that is not completely accurate, Fick still deserves a lot of credit for having clarified the situation in many ways and for having emphasized diosmosis through the membrane particles themselves in addition to diosmosis through capillary spaces.[72] I have already explained why we did not adopt the distinction between endosmosis and pore diffusion which Fick proposed, a distinction that would have corresponded to those different types of diosmotic passage. Lastly, Fick (1866:36) also was the first to point out the significance of the state of motion of salt and water molecules. But he overestimated this significance, since osmotic flow is basically caused by forces of attraction effective between heterogeneous particles, and the kinetic energy of the molecules is only a secondary factor in osmotic flow.

Having looked at views subsequent to and derived from Brücke's theory, let us turn to the divergent views held by Jolly (1849:145), Liebig (1848:51),[73] and Graham (1862:75). They agree that a variable boundary layer is not required. Also, the imbibed solution [*in the pore*] does not have to be variable across a layer taken parallel to the membrane surface. Such homogeneous imbibition could actually be attained if water or salt solution were absorbed only into the membrane particles. But since probably all permeable membranes contain other interstices as well and, specifically, because all membranes studied by the above-named authors have rather wide interstices, the theories assumed by Jolly, Liebig and Graham are inadequate because they disregard the variable boundary layer, a real and cooperating factor.

In other respects, the theoretical views of Jolly, Liebig and Graham diverge more in form than in principle. Speaking of principles, the theories of these researchers basically boil down to one fact -- depending on their mutual affinities, the membranes imbibe salt and water. When the membrane separates salt solution and water, the ratio of the opposed flows depends on the state of imbibition and attraction between salt and water particles. Thus a water flow that is predominantly unilateral will always be generated when, upon immersion in salt solution, a more dilute solution is imbibed. Obviously, it is quite immaterial whether Jolly speaks only of attraction or Liebig and Graham speak of chemical affinity. It must be emphasized, however, that Jolly does not refer to the contraction in salt solution of substances capable of swelling, while Liebig and Graham regard the very fact that there is unequal contraction on the two sides of a membrane, which forms a diaphragm that separates salt and water, as a driving force. Of course, such a state of contraction expresses the different salt content in material that is capable of swelling. But the state of contraction is not necessary for osmotic flow, since this flow can also take place in membranes that are not capable of swelling. Graham's assumption -- that diosmosis seems to reduce the unilateral osmotic water flow (Graham's osmosis), had long since been recognized as a logical consequence of Brücke's theory, which neither Liebig nor Graham had noticed.

The crystalloids, at least, diosmosed through all membranes used up to that time. That is probably the reason why the simplest case was not chosen as the starting point of the theory -- i.e., a case where the osmotically effective solute is unable to pass through the membrane. This is a given with regard to colloids with animal membrane and similarly acting membranes. Graham also had this type of behavior in mind when he studied osmosis, though he never correctly and adequately used this behavior to explain the process. Only when Traube discovered the precipitation membranes did it become absolutely necessary to take a closer look at the osmotic behavior of substances that do not diosmose. With these precipitation membranes, for the first time, membranes had been obtained that had the same type of structure and at the same time had osmotic properties, something no other membrane outside the organism had ever provided until that time.

Traube (1867:131) immediately dealt very clearly, in substance, with the formation and growth of precipitation membranes. He also tested different types of membranes for permeability to various substances, and in the process became aware that colloids produce relatively little water flow. But Traube (1867:147) was basically wrong in his theoretical explanation of actual diosmotic behavior.[74] He neglected the molecular forces effective between the membrane on the one hand and the solute and dissolving medium on the other hand. He formed the incorrect view that the membrane simply acted as a sieve and that, therefore, by studying whether different substances passed or not through a membrane, one could obtain a relative measure for the size of the molecules existing in the solution.[75] Another concomitant of his neglect of the molecular forces involved is an incorrect view of the osmotic effect. Osmotic effect is indeed (as Traube assumed) caused by attraction between the dissolving medium and the solute, but (and this Traube overlooked) its intensity depends basically on the composition of the diffusion zone.

17. Experimental Results.

The following is a compilation of the experiments for which only the results were given in the text of this monograph. All the series of experiments in Tables I through XV were done with copper ferrocyanide membranes; after Table XV, the type of membrane used is specifically listed.

Measurements of the osmotic water flow

The method I used was explained earlier (p. 11), and so I shall only state whatever is needed to explain the experiments in detail here. There is a note that indicates in each series of experiments whether the experiments were performed without membrane-formers (as was generally the case) or with them. The effective area of the precipitation membrane is also given when this was determined; that was done simply by calculating, from the diameter and the height, the area covered by the membrane. True, this measurement is only very approximate, but I found it completely adequate, since the thickness of the membrane and other factors influencing the water flow remained undetermined.

The readings for both thermometers, positioned at the upper and lower end of the cell (including the closing pieces), differed by 0.1° to 0.3° in the different series of experiments; yet special care was always taken that in a single series of comparative experiments this difference remained constant. For simplicity's sake only the mean of both temperatures is given in what follows.

Since in each series of experiments the temperature was constant, and the ratio of the osmotic water flows always alters only imperceptibly when temperatures are not far apart, the results obtained at temperatures that are not quite equal are commensurable when normalized. The experiments and the measurements of pressure were carried out in very diffuse light or in darkness.

Since in an experiment little water is introduced into the cell, the concentration of the content does not noticeably decrease and, in spite of simultaneous diosmosis, no decrease of the specific gravity of the solution was found for potassium nitrate. Therefore in the tables there were no further references to the determination of specific gravity.

At the end of a series of experiments a control experiment was done with the same solution as at the beginning, and if the intensity of the water flow had changed then (generally) the mean of both calculations was taken as the measure for comparison with the values found for other solutions. This procedure is not quite correct in itself, but the accuracy it offers (since differences are minor) is adequate for our measurements, especially since in comparative experiments with different solutions the experiments were often carried out in exactly reverse sequence. If the sequence does not coincide with the listing of experimental results enumerated from top to bottom, then the reverse direction in which the experiments were carried out is indicated by an arrow placed next to the table [*omitted, to simplify data presentation*].

In column (c) the concentration is given in percent by weight. The second column lists the duration of an experiment from which the values listed under (h), calculated per hour were obtained; these values indicate the increase in height of the column of fluid in the measuring tube expressed in millimeters. In the last column the mean values of column (h) are normalized as indicated. In part E, (e/c) designates the corresponding quotient (the inflow values reduced to a unit divided by concentration in percent by weight). Finally, column e/(c.s) lists the analogous quotient related to percent by volume, for c.s (s being the specific gravity of the solution) gives the concentration of a solution in volume percent [*note in part E that column s gives the specific gravity of the sucrose solutions at 17.5°C*]. Experiments A-D were all carried out with different cells.

I. Experiments with sucrose.

I

A

c (weight %)	time (hr.)	h (mm/hr)	h (normalized)
1	2.5	1.6	1.0
1	3.0	1.8	
2	1.5	3.6	2.1
6	1.0	10.1	5.9
10	0.75	19.8	11.6
20	0.75	43.4	25.5

(15.1°C, no membrane-formers, membrane area = 16.1 cm^2)

B

c (weight %)	time (hr.)	h (mm/hr)	h (normalized)
1	2.5	1.8	1.0
1	3.0	2.0	
2	2.5	3.5	1.8
6	1.0	10.7	5.6

(17.6°C, with membrane-formers, membrane area = 16.9 cm^2)

C

c (weight %)	time (hr.)	h (mm/hr)	h (normalized)
1	3.5	1.8	1.0
1	3.0	1.8	
6	1.5	10.5	5.8

(17.3°C, no membrane-formers)

D

2	2.0	2.0	1.0
2	3.0	2.1	
16	1.0	20.5	20.0
32	0.75	49.6	48.4

(15.2°C, no membrane-formers, membrane area = 15.1 cm^2)

E
(Summary: A - D)

c	e (mean values)	e/c	s	e/(c.s)
1	1	1	1.004	1
2	1.95	0.98	1.008	0.97
6	5.77	0.96	1.024	0.94
10	11.6	1.16	1.0404	1.11
16	20.0	1.25	1.0657	1.17
20	25.5	1.27	1.0832	1.17
32	48.4	1.54	1.1391	1.35

II. Experiments with gum arabic.

Gum arabic dried at 100°C was used to make a standard solution of a known concentration, from which solutions were made with the concentrations indicated below. The cells used were those with as thin a precipitation membrane as possible, and where membrane area was relatively large. Both experiments were performed without membrane-formers. From two experiments with the same cells, using 1% solutions, it was found that when the osmotic water flow of sucrose is set at 1.0, that of gum arabic = 0.14.

II

A

c (weight %)	time (hr.)	h (mm/hr)	h (normalized)
1	6	0.6	1
6	2.5	2.25	3.7
18	1	10.3	17.2

(15.1°C, membrane area = 17.1 cm^2)

B

18	6	10.2	15.7
6	2.5	2.3	3.5
1	5.5	0.65	1
18	1	10.4	-

(15.5°C, membrane area = 17.5 cm^2)

C - summary of A and B

c	e	e/c	s	e/(c.s)
1	1	1	1.004	1
6	3.6	0.6	1.025	0.58
18	16.45	0.91	1.078	0.84

III. Experiments with KNO_3.

III

A

c (weight %)	time (hr.)	h (mm/hr)	h (normalized)
1	2	4.6 }	1.0
1	2	4.2	
2	1.5	7.5	1.75
4	1	15.0	3.41

(13.9°C, with membrane-formers)

B

c (weight %)	time (hr.)	h (mm/hr)	h (normalized)
1	2	4.9 }	1.0
1	2	5.1	
2	1.5	8.2	1.64
4	1	16.5	3.3

(14.1°C, no membrane-formers, membrane area = 15.4 cm^2)

C

c (weight %)	time (hr.)	h (mm/hr)	h (normalized)
1	1.5	7.0 }	1.0
1	1.5	7.2	
2	1	14	1.97
4	1	25.1	3.53
8	0.75	45.9	6.46
18	0.5	83	11.69

(17.4°C, no membrane-formers, membrane area = 17.1 cm^2)

D - summary of A - C

c	e	e/c	s	e/(c.s)
1	1	1	1.006	0.99
2	1.79	0.89	1.013	0.88
4	3.41	0.85	1.025	0.83
8	6.46	0.81	1.051	0.77
18	11.69	0.66	1.123	0.59

IV. Experiment with K_2SO_4.

IV

c (weight %)	time (hr)	h (mm/hr)	e	e/c
1	1.5	7.0 }	1.0	1.0
1	1.5	7.0		
4	1	23.2	3.31	0.83

(15.8°C, no membrane-formers)

V. Experiments with 1% solutions of sucrose, K_2SO_4 and KNO_3.

Experiment B was begun and ended with potassium sulfate, and from the obtained values, 9.8 and 9.6, the listed mean value of 9.7mm was taken. -- With potassium sulfate and potassium nitrate, the experiment lasted 1.5 hours, with sucrose it lasted 2 and 2.5 hours.

V

A

solution	h (mm/hr)	e
sucrose	1.9 }	1.0
sucrose	1.9 }	
K_2SO_4	8.3	4.37
KNO_3	8.7	4.58

(15.7°C, no membrane-formers)

B

solution	h (mm/hr)	e
sucrose	2.2	1.0
K_2SO_4	9.7	4.41
KNO_3	10.2	4.64

(15.1°C, no membrane-formers)

(summary of A and B)

solution	h (mm/hr)	e
sucrose	-	1.0
K_2SO_4	-	4.39
KNO_3	-	4.61

Filtration under pressure

Regarding the method used, see p. 16. There also the corrections are listed that were used when the effective mean pressure listed in (d) was calculated from the mean level of the mercury column. The second column lists the time periods during which the mercury column in the pressure tube was observed to fall -- mercury levels are listed under (r) in millimeters. From the lowering of the mercury column, the amount of filtration was calculated with the help of the corresponding calibration values; the amount of filtration has been normalized to one hour and listed in column (f) expressed in mm^3. The quotient (f/d) and the normalization shown in column (m) appear in the headings. The two series of experiments that follow were carried out with different cells.

VI

A

d (cm Hg)	time (hr)	r (mm)	f (mm^3)	f/d	m (normalized)
112.2	2	9.3	10.3	0.0927	0.992
111.5	2	9.5	10.5	0.0942	1.009
71.3	3	9.5	6.54	0.0917	0.982
37.8	6	8.7	3.52	0.0934	1.000

(bath at 14°C, air near standpipe at 15.1° - 15.7°C, above, and at 14.7° - 15.2°C, below --the barometer fluctuated by only 1mm during the experiment --no membrane-formers, membrane area = $16.5cm^2$)

B

d (cm Hg)	time (hr)	r (mm)	f (mm^3)	f/d	m (normalized)
210.2	1.5	15.9	18.5	0.0880	1.023
208.0	1.5	15.0	17.5	0.0841	0.978
85.1	5	18.3	7.56	0.0888	1.033

(bath at 13.5°C, air near standpipe at 14.6° - 15°C, above, and at 14.2° - 14.7°C, below -- the barometer fluctuated by 0.5mm -- no membrane-formers, membrane area = $15.4cm^2$)

Table 6 on p. 71 combines the data given in columns(d) and (m) from A and B.

Pressure measurements

The method used and the calculation are discussed on p. 24; refer to that page for the meaning of v°, V, S and P [*V is in mm divisions of the manometer scale, while S and P are in cm Hg*]. In the following tables, I gave Brigg's logarithm of v°. As can be seen on p. 23, the temperature of V was measured by the thermometer whose mercury bulb was placed in the liquid contained in the bath at the mean height of the air-filled part of the manometer. The temperature indicated by this thermometer generally varied by 0.1° - 0.3°C from that of the thermometer placed lower down with its bulb near the lower end of the cell. Since, however, this difference was kept constant during the final decisive readings, and since minor temperature differences have no measurable effect on the osmotic pressure, it was not necessary to give specific temperature readings for the thermometer that was placed lower down. The temperature of the air enclosed in the manometer may also be simply regarded as the temperature of the cell contents.

The individual experiments within a series are always given in the same order as they were carried out.

VII. Pressures for sucrose solutions of different concentrations.

The two series, A and B, were done with different cells, but experiments within each individual series were done with the same cell. The separate experiments are listed in the sequence in which they were carried out.

The solution of membrane-former, as always, contained 0.1% potassium ferrocyanide and 0.09% copper nitrate (p. 27), except in experiment 2 where the concentration was twice as high, i.e. 0.2 and 0.18% respectively. In experiment 4, by mistake, a sucrose solution of unknown concentration, but with the usual potassium ferrocyanide concentration, had been used. The concentration of the solution used was determined to be 2.74% sucrose; it was determined optically (p. 31) and by calculating the specific gravity. After the end of the experiments, the specific gravity of the sucrose solution was seen to be unchanged, except in the experiment with the 6% solution where the 4th decimal had been lowered by 2, which, however, was not further taken into consideration.

Table 9 on p. 82 was combined from VII A and B by setting the pressure found for 1% sucrose solution (or the mean value of this pressure) equal to 1 in both A and B. Thus column three of Table 9 was obtained as the mean of these relative values normalized as above.

VII

A

No.	c (weight %)	V (mm)	temp. (°C)	S (cm)	P (cm)
1.	1	108.1	13.7	69.9	53.8
2.	1	107.7	13.6	70.9	53.2
3.	2	79.4	14.0	67.0	101.6
4.	2.74	61.6	13.5	65.1	151.8
5.	4	49.2	13.8	63.8	208.2
6.	6	36.3	14.7	62.4	307.5
7.	1	108.8	14.6	69.8	53.5

(membrane area = 17.1cm^2, log v° = 4.10489 for Nos. 1-4 and 4.10519 for Nos. 5-7)

B

	c (weight %)	V (mm)	temp. (°C)	S (cm)	P (cm)
	1	119.3	16.1	69.9	47.2
	6	42.1	15.4	63.1	267.9

(for both solutions, specific gravity was unchanged after use, log v° = 4.12026 for both)

VIII. Pressure for gum arabic.

The solutions used were prepared at the same time, from the same material and in exactly the same way as those used to determine the osmotic water flow (Table II). Both series of experiments, A and B, were performed with two different cells, and were done with membrane-formers in the usual way. Only in the experiments with 18% solution, where the results appear with asterisks, were the membrane-formers completely omitted. The specific gravity did not change in the course of an experiment. The experiments carried out with an open manometer have been designated as such.

VIII

A

Solution	c (weight %)	V (mm)	temp. (°C)	S (cm)	P (cm)
sucrose	1	111.6	15.2	70.6	52.6
gum arabic	1	168.2	16.1	75.9	7.2
gum arabic	6	142.3	16.1	72.0	26.3
gum arabic	18	75.9	15.7	64.3	119.7
gum arabic*	18	75.6	15.6	64.3	120.4
sucrose	1	112.6	16.6	70.0	52.7
	(with open manometer)				
sucrose	1	-	15.0	-	7.1
gum arabic	6	-	15.7	-	28.7

(membrane area = 16.9cm^2, log v° = 4.11492 for sucrose and 4.12089 for gum arabic)

B

Solution	c (weight %)	V (mm)	temp. (°C)	S (cm)	P (cm)
sucrose	1	117.0	15.8	71.3	48.1
gum arabic	6	143.6	15.9	71.4	24.6
gum arabic	18	74.6	15.5	65.7	118.9
gum arabic*	18	75.0	15.6	65.6	118.0
	(with open manometer)				
sucrose	1	-	15.4	-	6.7
gum arabic	6	-	15.3	-	24.0

(membrane area = 17.0 cm^2, log v° = 4.12089 for sucrose and 4.11492 for gum arabic)

Table 10 (p. 82) is put together from A and B in exactly the same way as Table 9 from A and B in Table VII.

If in A and B the pressure of 1% gum arabic solution is set equal to 1, then in A the pressure of 1% sucrose solution equals 7.41, and in B it equals 7.18. Thus the mean ratio of the pressure produced by 1% solutions is as follows: if gum arabic = 1, sucrose = 7.29 whereas if sucrose = 1, then gum arabic = 0.138.

IX. Experiments with sucrose, dextrin, K_2SO_4 and KNO_3.

These experiments were done to determine the pressure produced by 1% solutions of the above-named substances. For the diosmosing potassium nitrate, one experiment used a solution of 1.05%, while a second used a solution of 1.07%. The concentration was calculated at the end of the experiment from the specific gravity. The potassium sulfate solutions used to fill the cell contained 1% and 1.01% respectively of this salt. With the other solutes, specific gravity did not decrease in the course of the experiment. Where final concentration was not 1%, the pressure for the solution was calculated assuming proportionality between pressure and concentration, and the results were indicated by parentheses in the last column. Since the ratio of the two is actually different, but not exactly known, this calculation was in error. The error is probably minimal because of the small deviation from the 1% value.

The dextrin had been described as chemically pure by the manufacturer. But since I had not specifically purified it, I can only state that if glucose was present at all, it was at any rate present only minimally.

The osmotic water flow produced by a 4% sucrose solution was found to be 11.6, mm/hr, before the beginning of this series of experiments, immediately after the cell had been prepared, and was found to be 5.1 at the end of the experiments listed here. Thus it had decreased by more than half, probably largely because of blockage, although the pressure measured with sucrose solution had not undergone a change. The experiments were performed between March 2 and 14, 1876.

Table 8 (p. 76) was obtained by taking the mean values of the pressures determined for 1% solutions. This table also includes the relative pressure of gum arabic according to the value found in Table VIII; the pressure is thus 47.1 times 0.138 = 6.5cm.

IX

Solution	c (weight %)	V (mm)	temp. (°C)	S (cm)	P (cm)
sucrose	1.00	117.3	15.9	71.8	47.1
dextrin	1.00	152.2	15.6	75.0	16.6
KNO_3	0.98	58.2	15.8	64.7	174.9 (178.4)
K_2SO_4	0.98	55.1	16.1	64.6	188.8 (192.6)
KNO_3	1.01	58.6	16.1	63.3	175.0 (173.3)
K_2SO_4	1.00	54.9	15.5	62.1	191.9
sucrose	1.00	119.3	16.0	69.9	47.2

(membrane area = 17.0cm^2)

X. Experiments with KNO_3.

In this case, the cell was filled with solutions of 1, 2, and 4% and then the concentration, which decreased by diosmosis, was determined from the specific gravity. The concentration decreased in unequal amounts, and this is largely explained by the fact that the individual experiments were of different duration. This series of experiments was done between January 9 and 19, 1876. Table 11 (p. 84) was derived from the values below.

X

c (weight %)	V (mm)	temp. (°C)	S (cm)	P (cm)
0.80	67.4	13.2	67.6	130.4
1.43	46.9	12.9	65.7	218.5
3.3	26.7	13.0	62.7	436.8
0.86	62.1	12.6	66.9	147.5

Pressure with fluctuating temperatures

The individual series listed in Tables XI through XV were carried out by heating or cooling the cell or the whole bath, respectively, and the pressures corresponding to the temperatures were calculated in the usual manner. By determining the specific gravity of the solution here too, a possible change in concentration could be determined, when such a change had been caused by diosmosis in the course of a series of experiments.

XI. Experiments with 1% sucrose solution.

The constancy of specific gravity and the return of the pressure to the initial level (when the same temperature is reestablished) indicate that, in the course of the 5 to 12 days which a single series of experiments took, sucrose did not diosmose perceptibly.

XI

A (13.5° to 32°C)

log v°	V (mm)	temp. (C°)	S (cm)	P (cm)
4.15130	120.8	13.5	72.0	54.1
4.15130	125.6	32.0	71.6	54.4
4.15130	122.6	14.8	70.9	50.9

B (6.8° to 22°C)

log v°	V (mm)	temp. (C°)	S (cm)	P (cm)
4.10489	107.8	13.2	71.7	52.1
4.10489	107.8	22.0	72.8	54.8
4.10489	108.0	13.8	71.6	52.2
4.10489	108.1	6.8	70.2	50.5
4.10489	108.3	14.2	70.6	53.1

C (15.1° to 36°C)

4.11751	113.0	15.9	70.5	52.1
4.11751	116.3	36.0	70.8	56.7
4.11751	113.2	15.1	70.2	52.0

The results of this series of experiments, which were carried out with three different cells, have been put together in Table 12 (p. 86).

XII. Series of experiments with 14% solution of gum arabic.

Since this gum arabic solution was not prepared from the same material that was used in the experiments described in Tables II and VIII, the values obtained there and here are not comparable. This series of experiments was carried out between April 13 and 19, 1876.

XII

log v°	V (mm)	temp. (C°)	S (cm)	P (cm)
4.12059	98.3	13.3	70.9	69.9
4.12059	105.2	36.7	69.9	72.4
4.12059	100.7	13.3	68.9	68.6

XIII. Experiment with saturated potassium bitartrate solution (see p. 89).

XIII

log v°	V (mm)	temp. (C°)	S (cm)	P (cm)
4.12171	100.2	13.0	70.0	68.3
4.12171	79.7	29.2	68.0	115.8

XIV. Experiments with sodium potassium tartrate (see p. 91).

In A, the cell was filled with a solution containing 1% of the above salt. At the end of the experiment (after 8 days), the concentration had dropped to 0.94%.

In B, the solution in the cell contained 0.6% sodium potassium tartrate. The concentration at the end of the experiment (after 7 days) was not determined directly, but, as the pressure indicates, it cannot have changed much.

XIV

A

log v°	V (mm)	temp. (C°)	S (cm)	P (cm)
4.18477	69.2	13.3	65.8	147.6
4.18477	72.0	36.6	65.5	156.4

B

log v°	V (mm)	temp. (C°)	S (cm)	P (cm)
4.12171	85.4	12.4	70.4	91.6
4.12171	89.6	37.3	69.6	98.3
4.12171	87.3	14.2	69.5	90.0

XV. Sucrose - sodium chloride (see p. 92).

Quantities of both substances corresponding to the composition of the crystallizing double salt ($C_{12}H_{22}O_{11}$ + NaCl) were dissolved so that the solution contained one gram sucrose and 0.171g NaCl per 100g of solution. The experiment took three days; the specific gravity was not determined at the conclusion.

XV

log v°	V (mm)	temp. (C°)	S (cm)	P (cm)
4.11751	72.4	14.5	67.0	123.6
4.11751	76.0	37.9	67.1	129.2
4.11751	4.1	15.0	65.9	120.7

Experiments with prussiate of iron and calcium phosphate membranes

XVI. Measuring the pressure with prussiate of iron membrane for 1% solutions of sucrose.

The two experiments for which data are given below were done with two different cells. Regarding the concentration of the membrane-formers, see p. 29. The specific gravity of the solution had not changed during the experiment.

XVI

log v°	V (mm)	temp. (C°)	S (cm)	P (cm)	
4.11751	125.3	13.2	72.4	37.3	38.7
4.11751	122.3	13.9	72.4	40.2	

XVII. Measuring the pressure with calcium phosphate membrane for 1% sucrose solution.

The concentration of the membrane-former solution necessary for osmotic equilibrium was determined only relatively, i.e. with respect to solutions whose concentration had not been calculated exactly. Incidentally, for 0.1% calcium chloride almost 0.4% of ordinary sodium phosphate is required. The specific gravity of the solution in the cell did not change during the experiment.

With the same cell, a pressure of 3.8cm was found with an open manometer at 17.9°C for a 2% conglutin solution prepared with a small amount of potash, see p. 76.

XVII

log v°	V (mm)	temp. (C°)	S (cm)	P (cm)
4.12171	127.7	15.2	73.3	36.1

Experiments with parchment paper and animal membrane

The experiments were done with the apparatus described on p. 10, and pressures were measured with an open manometer. In order to have almost 6% solutions in the apparatus at the end of the experiment, correspondingly, a more concentrated solution was employed. The final concentration, as in other experiments, was calculated from the specific gravity and noted in column (c). I cannot give more comprehensive data, since in this case, at any rate, only very approximate values are what matter (incidentally, see p. 75). In both cases the area of the membrane was 5.3cm^2; the whole apparatus contained approximately 34 cc of solution.

XVIII. Series of experiments with parchment paper (rather dense and moderately thick).

XVIII

Solution	c (weight %)	P (cm)	temp. (°C)	P (for 6% solution)
gum arabic	6.08	18.0	24.1	17.7
sucrose	5.92	28.6	24.7	29.0
gelatin glue	5.83	20.7	23.3	21.3
KNO_3	6.00	20.4	24.1	20.4
gum arabic	6.18	18.7	24.3	18.1

XIX. Series of experiments with animal membrane (beef pericardium).

XIX

Solution	c (weight %)	P (cm)	temp. (°C)	P (for 6% solution)
gum arabic	6.01	14.2	21.4	14.2
sucrose	6.2	15.0	21.5	14.5
gelatin glue	5.81	15.1	21.1	15.4
KNO_3	5.73	8.5	21.3	8.9
gum arabic	5.9	12.1	21.6	12.3

The values calculated for 6% solutions in Tables XVIII and XIX are shown in Table 7 (p. 75). The osmotic effect for gelatin glue with a copper ferrocyanide membrane was measured with an open manometer. A 6% solution of the same gelatin glue was used to prepare the solution, as for the above experiments. The gum arabic employed here was the same material as that which served to determine the pressure with the copper ferrocyanide membrane.

Notes: Physical Part

1. See Traube (1867) - the first paper by Traube on this subject is Traube (1866) and he gives a summary of earlier, and some more recent, experiments in Traube (1874), for a reprint of which see Traube (1875).

2. The physiological section of this treatise will explain what is meant by plasma membrane.

3. The inner seal of the sealing-wax ring can best be checked by means of small mirrors of the type used by physicians, e.g., to examine the larynx.

4. The cells which I used at first, which I purchased from the warehouse of E. Marquardt and Co. in Bonn, were satisfactory even for producing internal membranes. Since the stock of these cells was quite small, and later shipments from the same factory were actually unusable, and since moreover I was at first unsuccessful in my attempts to produce surface membranes, it obviously took a great deal of time and effort before the technical difficulties were surmounted.

5. By chance, Kürschner (1842:57) also obtained such membranes. Attempting to monitor the passage of a liquid through a membrane, by means of a visible reaction, he separated solutions of potassium ferrocyanide and copper sulfate by means of an animal membrane. Since Kürschner did not at all recognize the significance of the precipitate that was formed in the membrane, no significance need to be attached to that experiment. The infiltration of diaphragms with lead chromate, performed by Brücke (1843:89), and with barium sulfate, done by Ludwig (1849:25), is valuable as an example of an experiment guided by a particular idea, but whose goal was not to form a precipitation membrane, nor would it have produced such a membrane if the membrane-formers listed here had been used. Incidentally, many of the so-called metallic trees, which appeared in the work of the alchemists, were formed of precipitation membranes. The iron tree of Glauber is an iron silicate produced from iron salt and waterglass [*sodium silicate*] (Kopp, 1847:149).

6. For the sake of brevity, I will speak of water as the outside liquid, though this is a dilute solution of the membrane-former.

7. All glass apparatus used was manufactured by Dr. Geissler in Bonn.

8. A somewhat modified construction, such as the one used by Regnault, (1847:329) would be best.

9. In actual fact, the solution contained in the cell always had a somewhat higher specific gravity than the outside solution. If this circumstance is neglected, however, the resulting error is too small to take into consideration.

10. Of course, copper sulfate could also be used in the appropriate concentration.

11. This inside solution was prepared with distilled water, while the outside solution is prepared with rainwater.

12. These experiments are presented in Section 17, experimental results, in Tables VIII A and B.

13. The method and margin of error in the determination of specific gravity are carefully discussed by Kohlrausch (1857).

14. See Wüllner (1870:Vol. II, 591). When I use a sodium light, my sucrose determinations with this apparatus are generally more accurate than 0.1%.

15. See Lothar Meyer (1872:277) and A. Naumann (1872:21 and 1873-75:293). The above writings explain to what extent there is reason to think that liquid and solid substances are compounds of molecules existing isolated in gas form. A number of chemical experiences point to the existence of complicated chains of atoms and molecules, as was recently emphasized by Zincke (1876:243), among others.

16. For a discussion of how compounds of molecules continue to exist in solutions, see Naumann (1876-77:477).

17. I cannot describe a definite difference between colloids and crystalloids, but I will keep this difference for the sake of convenience by calling a substance that diosmoses relatively easily through an animal membrane or parchment paper a crystalloid, as is the practice now, while I shall call one that diosmoses with relative difficulty through such a membrane a colloid.

18. Other authors give a somewhat different water content. Let me note in passing that it is generally known that the precipitate $Cu_2Fe(CN)_6$ obtained with potassium ferrocyanide always contains some bound potassium.

19. According to Graham -Otto (1863:1202), prussiate of iron dried at 100° contains 12 equivalents of water. For calcium phosphate, see p. 554.

20. We need not address the question to what extent the determining forces here are to be regarded as extending from the surface or the mass of the body.

21. The condensation of the layer is then modified, of course, but the total attraction exerted upon any arbitrary point within need not be reduced to zero, even where the capillary space is spherical. As mechanics teaches us, this is the case only for conditions which are not realized here, for the attracting and repelling forces involved, or which at least need not be given.

22. Cloetta (1851) performed similar experiments, see his dissertation, also see Fick (1866:31).

23. According to Liebig (1848:50) an analogous situation supposedly occurs when porcelain cells are dipped into salt solution. However, these experiments were not performed carefully enough for them to be regarded as conclusive.

24. Here, of course, I mean only such membranes as have a perceptible influence on the exchange of substances, and whose use does not produce the type of mixing as, say, in hydrodiffusion.

25. It might also be possible to differentiate osmosis through diaphragms capable of swelling from osmosis through diaphragms incapable of swelling. When a diaphragm swells up, the size of the interstices is accordingly altered.

26. Diatagmatic osmosis would be even less capable of yielding a relative measurement for molecular size even if amphitagmatic osmosis could be ruled out.

27. A numerical expression for the probability of collision, i.e. of the penetration of a substance passing through the membrane into the sphere of action of the tagmas, could be developed in analogous fashion to the work of Clausius (1867:260). Such an expression could refute the objections of Buys-Ballot, who tried to present the slow diffusion of gases as being incompatible with their assumed molecular movement. Also see Clerk-Maxwell (1876).

28. Experiment could decide whether the passage of gases through rubber, and through red-hot plates of certain metals, is caused exclusively by absorption of the gases, as Graham interprets it, or whether it is a mixed process of the type described above.

29. See Eckhard (1866:68)--Matteucci and Cima (1845).

30. To simplify things, let me refer to solutes as salt.

31. We must also consider the possibility that diosmosis might only be produced when a concentrated solution acts upon the membrane. Of course, this does not yet follow from the experiments with sucrose I have reported.

32. The constants for potassium = 1 are, potassium nitrate = 0.912, potassium sulfate = 0.703, sucrose = 0.314. The latter constant was determined by Voit, the others were recalculated according to Beilstein's experiments.

33. We compare here solutions measured in weight percent, which do not have exactly the same density. However, the above constants refer to solutions that contain equal amounts by weight per unit volume. The error that thus results, incidentally, is only minimal.

34. Note that potassium nitrate diosmoses in small quantities, whereby the osmotic water flow is somewhat lowered (though only little).

35. Graham (1862:11) claims that for gum arabic and sucrose the diffusion velocity is calculated as 2.8 to 5. But according to Hoppe-Seyler (1866:14) gum arabic (which is not a defined chemical substance) seems to diffuse much more slowly. For other colloids the slow diffusion, however, is known (Graham, 1862:17), but there exist no accurate determinations of the diffusion constant.

36. This was glue suitable for producing cells from tannic gelatin glue according to Traube's method. This glue can be made faster than by prolonged boiling if concentrated glue solutions are poured hot into closed glass tubes, and then heated for several hours at 120° to 130°C. Since glue is only suitable for making a cell when it contains a certain concentration of gelatinizing glue, an appropriate composition must be produced by adding gelatin later.

37. Liquid egg albumin as found in eggs leaves approximately 3% ash, Kühne (1868:553).

38. In addition to endosmotic experiments with already mentioned membranes, and apart from precipitation membranes, experiments have also been done with eggshells, stone plates, porcelain cells, plant leaves, wood lamellae, coagulated egg white, and other material.

39. Similarly, the fact that calcium chloride does not deliquesce does not imply that it exerts higher osmotic action than a substance that is not hygroscopic. The low force it takes to keep a number of colloids in solution (see Graham, 1862:69) only goes to show that it has less affinity for water.

40. It is open to question whether the volumes are exactly the same, as I shall show later.

41. Eckhard's objections (1866:91) must be approached carefully, because different porcelain cells are so dissimilar that their porosity is an important contributing factor.

42. Eckhard (1866:87) found that the intensity of the salt flow was affected when the latter moved into pericardium against a pressure of more than 8cm mercury.

43. Fick (1855:83,85) stresses two points that to him seem inconsistent with diosmosis through narrow spaces. The assumption, that the endosmotic equivalent must decrease considerably when, say, solid substances are added to the salt solution and thus the mobility of the particles is reduced, no longer holds true as soon as the diffusion zone (as is actually the case) is kept constant by relatively large molecular forces (see p. 50). The other view, that the endosmotic equivalent must decrease rapidly if very dilute solutions are chosen for comparison, neglects the fact that the boundary layer changes with the concentration of the adjoining salt solution. It is also not true for membranes capable of swelling up, for another reason, the diameters of narrow pores vary with the concentration.

44. We know that due to this force quantities of gas are exchanged through narrow pores in inverse ratio to the square roots of their densities.

45. Through a circular piece of membrane 6cm in diameter, 0.14 g water filtered under a water pressure of 227cm. The highest value given for another sample in the same table is 32.7 g water. The calculation was made assuming proportionality between the pressure and amount of outflow. This is not exactly true for animal membrane, yet still yields approximate values.

46. During diffusion in liquids not only the active force of the molecules, but also their mutual attraction is decisive. Where the latter disappears, as in the gases, the constant can be expressed as a function of molecular velocity, as Loschmidt (1870) showed. It is as exact as can be expected: $k = (e \cdot \mu_1 \cdot \mu_2)/N$, where e is a constant factor, N is the number of molecules per unit volume, μ_1 and μ_2 are, respectively, mean velocities of the two diffusing gases. The general rule that Sachsse (1874) tried to derive from Graham's experiments, which were not sufficiently accurate, was that the number of molecules diffusing from a salt solution decreased as molecular weight increased during the diffusion of a

liquid. This rule is not capable of allowing deeper insight into the molecular forces active here. We must particularly remember that the molecular weight derived in the usual manner does not determine, or at least does not always determine, the number of particles in solution. Take, for instance, the soluble iron oxide hydrate, which is presumably formed as a colloidal substance by aggregation of molecules into tagmas.

47. Naumann (1876-77:451, 480). The unequal coloration of a cobalt chloride solution directly shows the existence of unequally constituted compounds of molecules or molecules, in solution. Increase in volume when certain salts are dissolved is due to special circumstances.

48. Of course, Beilstein (1856:187) asserts that diffusion velocity increases faster than the difference in concentration of layers that adjoin each other; like Voit (1867:234), I do not feel this assertion is really substantiated by the experiments. Voit (1867:419) found that for sucrose the diffusion coefficient increased with increasing concentration, but feels he can ascribe this to the fact that the saccharimeter data were too low owing to the diffusion flow; but he never proves that this error actually causes the entire deviation, and not just part of it.

49. Fick (1857:325) leaned toward this assumption as regards collodion membranes.

50. If we start with the assumption that the boundary layer keeps the same composition for solutions of different concentrations (which, of course, is probably not so), we could imagine that the boundary layer yields a constant water flow (c). Then, when it adjoins on a more concentrated solution, a diffusion zone would have to be created here, and the water flow would be expressed by the formula $w = c + u \cdot n$ if the output of the diffusion zone in question increased proportionally to the concentration. Then, the two unknowns c and u can be determined from two equations. However, the assumptions made here are not true, which is why the two quantities in question turn out not to be equal if they are derived from the outputs of pairs of two solutions at different concentrations.

51. The diffusion of the most diffusible salt alters least; occasionally a slight increase is seen, very often there is a certain decrease; the decrease, however, is always less than for the salt which is less diffusible.

52. On these, see the work quoted on p. 216 in Baranetzky (1872).

53. The apparatus I used consisted of a glass tube, one end of which was closed by the membrane, while the other, the tapered end, served to receive the measuring tube (see p. 10). It was always adjusted in such a way that a slight overpressure existed inside the cell, thus somewhat stretching the membrane right from the beginning. Since the pressure inside increased with the inflow, and slight pressure already causes somewhat considerable filtration through parchment paper, the values obtained for the osmotic inflow cannot have turned out very accurate, but must have been too low.

54. This was conglutin produced from Lupinus by Ritthausen.

55. See this work, p. 52.

56. It is conceivable that suction and pressure of equal intensity might not be able to produce the same water flow. If water is suctioned through a tube, the velocity of the flow cannot exceed a value, dependent on prevailing conditions, if the continuity of the flowing water is not to be interrupted. At this value, a certain role is played by cohesion, which varies as water is compressed in the boundary layer. Still, as I shall state later, this factor seems not to play a significant role in pressure; once the pressure is reached, of course, the two opposing water flows balance each other.

57. In a preliminary paper I stated that osmotic pressure fell with rising temperature (Pfeffer, 1875a and 1875b). During the first 3 experiments, which I carried out at different temperatures, it is true I had each time observed a pressure decrease, and my assumption was based on these experiments. The different pressures must, however, have been the result of some damage to the cell. Moreover, as I acquired more experience and practice in preparing and working with the cells, I

encountered only one other such instance, under similar conditions to those in which I had previously observed lowering of the pressure. Here, the cause for the lowered pressure was also found to be damage to the precipitation membrane. Actually, the lowering of pressure we are discussing here was the result of a technical difficulty that had not yet been overcome. By going into more detail I could show how it was almost inevitable that I obtained such a result because I prepared the cells somewhat differently than I do now and, as experience later showed, inadequately. Still, I do not want to give excuses here, but to point out emphatically that my former assertion that pressure is lowered when temperature rises, due to a lessening of filtration resistance, is actually incorrect. Obviously, all the other inferences based on that assumption are untenable. Let me add that when I spoke of correlations between pressure and filtration resistance, I naturally always had in mind a membrane of the same thickness and of otherwise equal properties. However, there is no more need to speak about the fact that the osmotic behavior of a membrane was characterized, inadequately, by the amount of filtration resistance. In the preliminary paper, a misprint gave a pressure for a 2% sucrose solution. That pressure was actually produced by a 3% sucrose solution.

58. As temperature rises, the volume remains approximately constant, while of course the concentration (estimated in terms of weight percent) increases somewhat because of the quantity of fluid that filters out of the cells. Still, this change, caused by the expansion of the fluid, is too small to have to be taken into account.

59. Here the temperature of the cell content and the surrounding medium is identical. But when both sides of the membrane are kept at different temperatures then, as a result of this difference, a water flow, though only a weak one, might take place for analogous reasons as those underlying the so-called thermodiffusion of gases, i.e., the passage of gas through a diaphragm from the colder to the warmer side of the membrane (Feddersen, 1873). Neumann (1872) carried out theoretical investigations. A unilateral water flow can also be produced by electric currents, so-called electric osmosis. The only thing the latter has in common with

our osmosis, by the way, is the fact that the influence of the wall is necessary to produce the phenomenon, and that what is involved is not just the simple mechanical effect of the electric current (Wüllner, 1870: vol.IV, 602 and 640).

60. Eckhard (1866:78) tried to establish such a formula following the model of Lagrange's interpolation formula.

61. Schmidt (1856) found that the amount of filtration for animal membrane increased with increasing temperature, in the same way as described by Poiseuille for glass capillaries, using an empirical formula with a trinomial factor (Wüllner, 1870: vol. I, 294). This formula, by the way, most probably is hardly applicable to precipitation membranes, and in any case gives only a moderate approximation. Since, according to Meyer's formula (see p. 72), the amount of outflow (as Hagen found also) must reach a maximum for a certain temperature, since the factors represented by η and ρ do not change in the same ratio with temperature.

62. The older studies of Ludwig (1849:9) do not agree with this.

63. Dibbit (see Naumann, 1876-77:547) calculated the extractable quantity of ammonia; here, precisely because of this extraction, the reaction had to progress further, as Berthollet (1803) had already stated, that an otherwise only partial reaction could become total when one of the products of the decomposition is withdrawn. For the same reason, experiments in which -- through shaking out with ether etc. (Berthelot and St. Martin (1872), and others) or through diffusion and diosmosis (Graham, and others), or in some other way -- a separation is produced can only indicate if dissociation takes place at all, but cannot yield a quantitative measure for dissociation.

64. Not totally in agreement with it are Graham's experiments. Graham concludes, on the basis of diffusing quantities, that double salts do not decompose when they dissolve, but that on the contrary the components as they dissolve do not immediately combine into a double salt. These statements, which are unlikely per se, have never been specifically checked

since (Graham, 1851:84).

65. Wiedemann elegantly derives the amount of dissociated iron oxide from the unequal molecular magnetism of iron oxide in combination with acid and colloidal iron oxide (Naumann, 1876-77:549).

66. Insofar as such fluctuations cannot be used to decide certain questions.

67. Since I want to sketch the historical development of our knowledge of osmosis, I shall naturally consider only those works that truly helped promote research in a direction discussed in this paper.

68. Vierordt (1846) compiled the older literature on the subject up to and including Brücke, to which I refer the reader. Jagielski (1859) treats the literature up to 1859 in detail, but not very critically.

69. Dutrochet (1837) -- My quotations refer to the Brussels edition. -- Vierordt's paper lists Dutrochet's earlier publications.

70. I have quoted the work of the other authors repeatedly. For example, other studies were done by Cloetta, Buchheim, Adrian, Hoffman, Harzer, Schumacher and others.

71. This was also confirmed by Cloetta (1851).

72. Fick (1857:296) -- Responding to a few objections by Eckhard (see p.57), Schumacher (1861) offers no theoretical contributions that need special mention here.

73. Liebig's discussion of the osmotic state of motion lacks the kind of clarity usually found in this great scholar. I hope, however, that I correctly understood Liebig's view, since I followed specifically what Liebig himself essentially repeated in a postscript to Graham's work.

74. The total neglect of the molecular forces that extend from the membrane is all the more surprising because, in a footnote, Traube (1867:150) points out the behavior of India rubber which, when compared to the diosmotic

properties of animal membrane, demonstrates the influence of the membrane so convincingly.

75. Incidentally, another reason why the molecular weight and molecular size of solutes (Traube incorrectly says atomic weight and atomic size) are not in the same ratio for different solutes is because molecules can be aggregated into tagmas.

II.

Physiological Part.

18. The Plasma Membrane.

Since Nägeli (1855:5) pointed out the importance of the diosmotic properties of the protoplasm, it has always been assumed, to the best of my knowledge, that the entire protoplasm body *[Protoplasmakörper]* --and not, for instance, a part of it --is important for diosmosis. But, as can be inferred from known facts, a peripheral layer must determine whether a solute is or is not absorbed. For if a solute finds its way through the relatively static peripheral layer into flowing protoplasm, in which even solid substances are mechanically tossed about, it must necessarily be distributed in the protoplasm. Obviously, we assume here that chemical reactions or other processes do not disturb the distribution process. Since Traube's studies on precipitation membranes and their osmotic behavior, the non-absorption of certain solutes, which easily diosmose through membranes used in earlier experiments, is no longer thought to be an exclusive property of the protoplasm body.

Qualitatively, the physical structure of the peripheral layer of the protoplasm must be the decisive factor for whether a substance will be absorbed or not. There have been attempts to explain this ability to exclude by attributing it to processes related to activities of life *[Lebensthatigkeit]*. The fatal error in such attempts is that dead and living protoplasm share all the diosmotic properties that have been studied to date. But since even easily diosmosing crystalloids do not penetrate this water-imbibing peripheral layer, the particles that constitute this layer must necessarily be arranged in a dense layer, because the diameter of the interstices could at most be equal to the diameter of the effective sphere of the particles plus the diameter of a crystalloid molecule. This requirement is imperative, assuming we do not want to break with the principle of causality, in order to prevent the passage of substances, even if the intended penetration be with some energy because of molecular attractions between the solute and water (and possibly also particles of the peripheral layer). However, by such attractions, it is just as imperative that an osmotic pressure be produced, and this pressure is determined by the physical structure of the peripheral layer and the properties of the constitutent particles.

With respect to diosmotic exhange and diosmotic pressure, it is irrelevant whether, according to its state of cohesion, the peripheral layer of the protoplasm, in itself, is to be regarded as a membrane or not. The thickness of this layer also can have significance only for the amount of diosmotic exchange, but not for diosmotic pressure (p. 79). It is quite another matter whether this peripheral layer is a membrane or not. But in no case is it possible that this diosmotically decisive layer is only a so-called liquid film *[Flussigkeitshautchen]*, and it would of course not be one even if it was formed as a result of altered molecular action on the free surface of a liquid body.

The diosmotically decisive layer may, as a matter of fact, be very thin. This is shown by the sometimes minimal thickness of the peripheral covering of the flowing protoplasm and of the corresponding covering of protoplasm that is not conspicuously in motion. In theory, the diosmotically decisive layer could, of course, consist of only one layer of the smallest particles, and direct observation shows only that dissolved dyes do not penetrate even the outermost parts of the protoplasm covering. In such circumstances, it is not possible to say, on the basis of the absorption or the non-absorption of solutes from the outside environment, whether the hyaline border layer [*hyaliner Saum*], the surface layer *[Hautschicht]*, has the same diosmotic properties in all its concentric layers. This must appear unlikely at least in cases like the myxomycetes. There the surface layer has greater thickness and microscopic observation shows that its density decreases from the outside toward the inside. Thus (all other things being equal) it can be expected that a decrease in the density means a decrease in the diosmotic exclusion capacity. True, I am unable to prove this exactly, yet certain diosmotic observations in myxomycetes also support my idea. I am reluctant to speak of them here because they were not pursued with sufficient exactitude. For my physiological purposes, the thickness of the layer that regulates the absorption of a substance was not at all important.

One may call the entire hyaline covering of the protoplasm body the surface layer *[Hautschicht]*, surface plasma *[Hautplasma]* or, perhaps best of all, hyaloplasm, *[Hyaloplasma]*. If so, at those places where this forms a thicker layer, there is most probably only an outer zone that is decisive for the diosmotic processes, which we observe in the protoplasm. To express this, I have decided to call this diosmotically decisive layer "plasma membrane" *["Plasmahaut" or "Plasmamembran"]*, and it is obviously possible that perhaps the whole hyaloplasm is a "plasma membrane," so that both terms become identical. Like Strasburger (1876a:286).[1] I shall call the granular protoplasm the granular plasma *[Körnerplasma]*, and to express the visible differentiation with one word, I prefer to call the hyaline covering "hyaloplasm." Incidentally, the plasma membrane is part of the protoplasm body and the membrane-like covering of a body that does not belong to the protoplasm (e.g., of drops of tannic acid present in cell sap) will not have to be called plasma membrane even though it were to be physically and chemically identical with a plasma membrane.

The above differentiation of a plasma membrane was made only with regard to diosmotic behavior, and is perhaps morphologically not necessary. It is just that, since I had to have a precise term for the diosmotically decisive layer, I found myself in the unpleasant position of either taking the term hyaloplasm, surface layer or primordial utricle *[Primordialschlauch]* in a narrower sense,[2] or of giving a special name to the layer in question; the latter alternative finally seemed preferable. Incidentally, I regard even the term "plasma membrane" as a stopgap measure, and would be glad to stop using it as soon as a knowledge of the structure and properties of the hyaloplasm would allow me to. Besides, it is not likely that the plasma membrane is clearly delineated against the inner layers of the hyaloplasm. The conditions for the formation of the plasma membrane exist, under certain circumstances, no matter which causes are decisive, wherever protoplasm masses offer a free surface. But on the other hand the particles of the plasma membrane are distributed in the protoplasm when the plasma membrane is surrounded by protoplasm on all sides, and thus two opposing processes are at work --one working to form, the other to destroy, the plasma membrane. As a result of these antagonistic processes, the plasma membrane limits may be undefined on the inside. In fact, it must generally be so whenever the distribution of the membrane particles in the protoplasm is preceeded by swelling of the protoplasm.[3] Then it would be expected that density would decrease towards the interior and the entire layer of hyaloplasm might be produced in this way. I emphasize that I do not claim all hyaloplasm must be produced this way; on the contrary, I have good reason to believe that the sometimes relatively thick layers of hyaline protoplasm in myxomycetes owe their formation to another cause.

As long as formation material is there and the conditions for formation (growth) are given, the resultant of the membrane-expanding forces cannot explain the cohesion of the isolated plasma membrane. In the same way, such forces cannot reveal the strength of a precipitation membrane that grows in area as a result of expansion. However, it is true that the skin-like layer around protoplasm, under normal circumstances, behaves somewhat like a viscous body when subjected to pressure and pulling. This consistency need not be true of the plasma membrane, which is conceived of as separate, because increase in surface through growth and decrease in surface through dissolution are possible (to put it briefly). Indeed I shall show how, after abolition of the ability to grow, the protoplasm (which is not coagulated, however) can be preserved, covered with a resistant membrane that has the same diosmotic properties as the living protoplasm had previously. Only this membrane prevents the penetration of solutes into the dead protoplasm. We shall later consider the reasons to regard the covering peripheral layer, if conceived of as separate --a layer which covers living protoplasm --as a resistant membrane. Let me also point out that the interior parts of the hyaloplasm do deviate in density and diosmotic properties from the outer layer, but might nevertheless have the consistency of what is unquestionably a membrane. This indicates how inadequate and makeshift is our provisional differentiation of the plasma membrane. Of course, my intention was never to do a special study of the structure of the protoplasm body and its parts. I intended only to pursue this structure as far as seemed absolutely necessary to understand the osmotic processes. Therefore I will omit certain incomplete observations which, if pursued with more attention to detail, might help explain the constitution of certain zones in the hyaloplasm.[4]

Until now, discussions of the properties of the peripheral covering of the protoplasm have never paid attention to the diosmotic behavior and ability to grow of this covering. True, under normal circumstances the cohesive quality of a plasma membrane not capable of growth (thought of as being separate) can be disregarded because, under such circumstances, conditions for growth are present. And yet the cohesive quality of those peripheral layers that are capable of growth, and those which are not, should of course be differentiated. As Nägeli (1855:19, Nägeli and Schwendener, 1867:403) correctly points out, under normal circumstances the hyaloplasm (and thus the plasma membrane as well) behaves like viscous slime. Those who claim that the peripheral layers, under these circumstances, are far firmer and more resistant, i.e., those who regard it as a resistant membrane, contradict actual observations. Hitherto all discussions refer to the fact that there is ability to grow, but the implication of this fact for the resistance of the peripheral layers of the protoplasm was overlooked. After the above remarks, I feel I do not specifically need to explain the views of different authors who subscribe to Mohl's,[5] or Pringsheim's (1854:5), view regarding the aggregate state of the hyaloplasm, or who take a middle position between these two views. Mohl tended to regard the hyaloplasm, the primordial utricle, as a membrane, while Pringsheim viewed the hyaloplasm, which he called the surface layer, as a slimy mass. Some authors (M. Schultze, Kühne, Hofmeister) simply described the hyaloplasm as the outer denser layer of the protoplasm, and left the reader in doubt about the state of aggregation. Incidentally, the dispute whether the hyaloplasm was a membrane or not was pointless, as Nägeli (1855:19) pointed out, since no definition of "membrane" was given and this term can be interpreted in different ways.

Mohl and Pringsheim had in mind only the boundary surface of the protoplasm body, whereas,[6] Hartig regarded the protoplasm (the ptychode sap) as a body outwardly and inwardly bounded by a membrane (the pytchode). True, Hartig had incorrect ideas on the further significance of the ptychode. Nägeli (1855:9,10 and 1867:552) then, very generally, stated that wherever cell contents rich in proteins --and thus protoplasm as well --come in contact with other aqueous media, a membrane-like layer forms on the entire contact surface by a process comparable to the coagulation of proteins. As a specific case in point, protoplasm, too, is thus separated from cell sap by a membrane-like layer. This view seems to have been partially accepted subsequently (Hofmeister, 1867) and was energetically defended by Hanstein (1870) in more recent times. Incidentally, the delineation on all sides of the protoplasm body that contacts aqueous media, by the diosmotically decisive plasma membrane, is immediately shown by the diosmotic behavior and, specifically, by the behavior toward dissolved dyes.[7]

As I just stated, Nägeli observed that generally a membrane-like layer forms around protoplasm masses in water. He also observed that living protoplasm bodies and any type of small vacuoles formed out of protoplasm similarly do not allow dyes to diosmose. In both cases, the protoplasm remains unstained and an exchange of dyes cannot be noticed at all, both when these are dissolved in the outer medium and when they are dissolved in the fluid that is enclosed by the protoplasm.

If one wishes to work with vacuoles, these are best formed by crushing protoplasm masses in suitable solutions of sucrose or other indifferent substances in order to prevent the extensive enlargement and final bursting of the vacuole, which occurs easily in pure water through the osmotic action of the substances contained therein. If protoplasm is crushed in a solution containing dye, vacuoles containing the dye are obtained now and then. I have often tested vacuoles in 2%-10% sucrose solutions for their diosmotic behavior, particularly some produced from the protoplasm of Vaucheria geminata Walz,[8] from the root hairs of Hydrocharis morsus ranae, from beetroot and from the plasmodium of Aethalium septicum. At times vacuoles continued to exist for eight days in intensively stained solution kept at the same concentration level, and no perceptible trace of dye penetrated the protoplasm. This could be conclusively judged when the dye solution was quickly replaced by sucrose solution of the same concentration level, and no perceptible trace of dye penetrated the protoplasm.

Eventually, the vacuoles are destroyed and when they collapse, which always happens suddenly, dye immediately enters or leaves and is rapidly accumulated in the protoplasm, as is known.

I experimented in the manner described with water soluble aniline blue, carmine, hematoxylin (dissolved by means of a trace of ammonnia) and pigmented fruit juices, and noticed no penetration of a trace of color either in the vacuoles or in the living protoplasm. In other respects, too, the diosmotic behavior of both appeared to be identical (an account follows later). Vacuoles that were surrounded by as thin a layer of protoplasm as possible, and vacuoles that had been formed from hyaloplasm or granular plasma also showed the same type of diosmotic behavior.

Finally, in isolated masses of granular plasma the hyaline covering forms in the usual way, but I have not adequately tried to ascertain whether such masses, formed solely out of granular plasma, can also be viable. This probably seems likely for myxomycetes, but it cannot be asserted as a matter of course. The healing of injured areas, which is well-known in the case of Vaucheria and other material,[9] is produced, partly, when the edges of the wound approach each other. Even if hyaloplasm be partially formed from granular plasma, the granular plasma and hyaloplasm (which was present before the injury) still remained together in the living material, after the injury.

For my physiological purposes it was enough to know that plasma membrane is always formed on the free surface of protoplasm masses immersed in aqueous medium, and for that reason I did not pursue all aspects of questions that suggested themselves regarding its formation. Precisely because the plasma membrane (also hyaloplasm) only forms on the free surface, a free surface necessarily becomes a causative factor in the formation of a plasma membrane. To begin with, we would need to decide whether, for this to happen, an external influence is required or whether, independently of such an influence, conditions for the formation of the plasma membrane are provided by and in the free surface. Along this line, the behavior in water and also in other media could only be established if the plasma membrane was not formed without the influence of these media. For certain cases there is no proof that a plasma membrane actually exists. Thus masses of protoplasm in fatty oil, and in air,[10] are covered with a hyaline border layer but, at the least, if the diosmotic properties are not determined it will not be possible to state positively that plasma membrane has also been formed there. True, probability speaks for this possibility and, if not, then at least hyaloplasm has been formed on the free surface. The question posed above cannot be decided by that alone, since there were additional external influences involved. We must not forget that not only evaporation, for example, and processes related to it, but also the accumulation of a thin sheet of water, say between the oil and the protoplasm, might be active factors. It is because of this and other circumstances that the question at issue is difficult to decide unless, under certain conditions, the plasma membrane (or the hyaloplasm) does not form.

Be that as it may, the hyaline covering of the protoplasm

cannot, as Hofmeister (1867:3) would have it, be the directly visible expression of the capacity of fluid bodies to surround themselves with a surface of altered density. It is true that the particles on the free surface of a liquid are in a different state of density and mobility than the particles in the interior of the liquid. The premise, however, on which the theories of Laplace and Poisson are based--that the decisive molecular forces are effective only at a minimal distance--is incompatible with the above view, since the hyaline covering at any rate attains measurable thickness. Let me add that according to the theory of Laplace and Poisson compression must take place on the adhesion surface, where molecular action between the fluid and solid predominates.[11] While in the opposite case, i.e., on the free surface of liquids, an unmeasurable thin layer of lesser density must be formed.[12] It is true that the latter has not yet been experimentally proved, however nothing contradicts this corollary of the theory and, at any rate, it is not possible to assume a denser surface for the protoplasm on the basis of a property of liquid bodies, according to which, in fact, a less dense surface is to be expected. Incidentally, it is quite easy to understand that the capacity of the peripheral layer of protoplasmatic masses for diosmotic exclusion cannot be explained by a layer altered, in a similar way, as the so-called liquid film. I feel it is not necessary to offer special supporting evidence here.[13]

It is a totally different question whether the altered molecular action on the surface boundary of protoplasmatic masses can lead to the formation of plasma membrane or hyaloplasm. There is no lack of examples where, for instance, the excretion of certain substances is caused by means of a minimal impetus. We will have to be careful to withhold our judgment as long as nothing definite is known about the properties of the material that forms the plasma membrane or hyaloplasm, all the more so since the behavior of albuminous substances is still a mystery in some respects. Moreover, even on the surface of completely pure and homogeneous liquids, under certain circumstances, phenomena occur that could not be explained adequately until now.[14]

As things are now, only experiments can decide whether altered molecular activity results in the formation of plasma membrane or hyaloplasm at the surface, or if some kind of external influences are a necessary contributing factor. The latter probably seems more likely and it would be possible to give a whole series of probable reasons, but I shall not engage in a discussion that could not lead to a definitive conclusion. If, nevertheless, I deal with the formation of plasma membrane in water on the assumption that external influences are necessary, I do so with the awareness that my final conclusions collapse if my assumption is not valid. Yet, I do not believe I should keep to myself the observations I have available, because they could probably provide some useful material even if altered molecular activity at the surface of the protoplasm is the only decisive factor. Incidentally, what follows should be considered only as a fragment, because the formation of the peripheral covering of the protoplasm under other conditions than in aqueous media is not discussed.[15]

In order for a plasma membrane (and for the hyaloplasm) to form, either particles that are already insoluble must be brought together or solutes must be precipitated from their solution. The latter seems necessary to allow for the insertion of new particles during growth by intussusception. This is most extensively possible when, for example, the hyaline layer that covered the protoplasm, surrounding a vacuole, increased its surface area more than 40 times while its thickness did not perceptibly decrease. I shall show that, with certain precipitates in the protoplasm, the ability of the plasma membrane to grow disappears. If, according to this, we must postulate the existence of a solution of the membrane-former in the protoplasm, this obviously does not exclude the possibility that, to form a plasma membrane, already undissolved particles are brought together. These might be cemented together, so to speak, by the precipitation of previously dissolved particles. Suppose that plasma membrane were to form from less dense layers of hyaloplasm through simple compression, then this case, too, is already included under the above viewpoints. Acutally it does seem likely that something of that kind takes place, which of course could only be advantageous for the rapid growth of the surface of the plasma membrane. Indeed, I have observed several times that when the surface increased very rapidly, the thickness of the hyaline layer perceptibly decreased, but regained its former thickness after some time.

As soon as any type of membrane-like layer is formed by precipitation of a substance from a solution, this can be regarded as a precipitation membrane in accordance with the meaning of the word. We would, however, not be dealing with a precipitation membrane if only undissolved particles collected together. If this process and precipitation occur simultaneously, we still may speak of a precipitation membrane insofar as the precipitation is an integrating factor. At present I see no need to introduce a special term in this type of case. When we use the term "precipitation membrane," it is a matter of complete indifference how precipitation occurs and what qualities the material possesses. That is why the considerations regarding the plasma membrane, which I discussed above, are valid even in a case where, for example, the plasma membrane is formed at the surface layer as a result of altered molecular activity.[16]

If we make the above assumption, then the plasma membrane (or the hyaloplasm) must be formed by the action of water on the protoplasm since, in fact, pure water alone suffices to produce the plasma membrane. Let us disregard the case where the association of undissolved particles probably contributes to the formation of the membrane. Let us now focus specifically on the precipitation of the dissolved membrane-former. There are two possibilities; either this precipitation will be the result of dilution with water or of the withdrawl of a solvent. When dehydrating, but otherwise neutral, solutions are used, it is shown that the withdrawal of a solvent can in itself cause a membrane to form. For if protoplasm masses are crushed in sucrose solution that is sufficiently dehydrating (I used solutions with 10% - 30% concentration), then dilution with water at the contact surface is out of the question. Yet, a hyaline covering forms nonetheless, and the behavior towards dyes shows the existence of a plasma membrane, which also increases its surface area normally (i.e. without breaks in the continuity), if a more concentrated sucrose solution is replaced by a more dilute one. Accordingly, we may probably say that the withdrawal of the solvent alone suffices for precipitation and membrane formation, but this does not exclude the possibility that in a smilar manner the simple dilution of the solution of the membrane-former already causes this to be precipitated and a plasma membrane to form.

Traube,[17] already, has showed how a precipitation membrane can be formed by simple contact with water. A more concentrated, but not a dilute solution of tannic acid dissolves tannic gelatin glue. When a drop of such a solution is placed in water, tannic gelatin glue is therefore precipitated at the contact surface as a membrane. Here dilution with water actually produces precipitation. Experiments would have to decide whether in this specific case, too, the withdrawal of the solvent, the tannic acid, can alone cause a membrane to form. Incidentally, it would not be difficult to cause a membrane to form in this way if one were to set out to prepare suitable solutions.

Special experiments were needed to make sure that pure water caused the precipitation of the dissolved membrane-former. To begin with, it is easy to ascertain that the plasma membrane continues to grow normally if the water contains no solid substances except for sucrose and dextrin. One carefully rinses vacuoles with pure solutions of sucrose and dextrin, and then causes the volume of the vacuole to increase by diluting these solutions. This increase in size, brought about by the osmotic action of the dissolved substances (in the vacuole), is in itself a proof of the existence of the plasma membrane, which is also revealed by the fact that it keeps harmless dyes from entering. A very similar type of behavior was observed when the sucrose solution did not contain oxygen and carbonic acid.[18] To carry out these experiments, I placed individual vacuoles, which were immersed in a small drop of pure sucrose solution, in the slightly enlarged space of a small glass tube. Then for several hours a stream of pure hydrogen gas, specifically containing no oxygen or carbon dioxide, was passed through the tube. In a second, slightly enlarged, space there was a very small amount of water, which could be made to contact the sucrose solution by tilting the glass tube. Microscopic observation produced the results previously described.

In all probability the membrane-former is a protein, as I intend to demonstrate; but I cannot say of what type the solvent is. On the basis of facts that I will discuss later, it is possible to state with certainty only one thing, neither membrane-formers nor the solvent diosmose to a significant degree through the plasma membrane.

What I have said above shows that it is very possible that the plasma membrane is formed by contact with pure water; but, as I explained, it is not possible to judge whether our final conclusions on the subject are correct until certain other questions have been settled. Very generally, I think it is possible to say that the plasma membrane is formed or grows when the solution of the membrane-former is decomposed on the surface of a protoplasm body.

How the plasma membrane is formed is of no great significance for the following investigations and inferences. These will show to what extent there are reasons to think that the diosmotically decisive layer possesses the aggregate state of a membrane.

Sections of beet were immersed in 20% sucrose solution under a cover glass; the concentration of the solution was continually kept constant. Four to five days later there were still quite a number of cells in which the contracted protoplasm enclosed the cell sap, which was apparently as intensely pigmented as at the beginning of the experiment. When the sucrose solution was quite gradually diluted, the ability of the protoplasm to expand and to lie against the cell wall was now partially or totally abolished. When the sucrose solution was diluted, the pigment in some cells rapidly crossed the protoplasm covering to enter the surrounding medium even before the contracted body had noticeably increased in volume. In other cells, the pigment moved out after the protoplasm body had first become slightly or somewhat more expanded. What determines this behavior, which was quite analogously observed with other cells also, is that growth material for the plasma membrane is absent, or present only in small quantities. Therefore, this plasma membrane, which gives rise to the diosmotic properties, is burst immediately, or after a certain growth in area, by the rising osmotic pressure of the contents.

Completely analogous phenomena were observed with the contents of petal cells of Pulmonaria officinalis, which had been contracted by a 14% sucrose solution, and also in vacuoles formed from the protoplasm of Vaucheria or Aethalium in 3% or 6% sucrose solution stained with water-soluble aniline blue. Here the fact that the dye entered the unstained protoplasm and previously unstained vacuole fluid indicated that the plasma membrane had burst.

The ability of the plasma membrane, and thus the entire enclosed protoplasm, to expand is more rapidly destroyed by hydrochloric acid, which I used in the following concentration, one drop of commercial hydrochloric acid in 15 to 35cc of the medium used. If water thus acidified is brought into contact with beet cells, the petals of Pulmonaria or other suitable material, the almost instantaneous reddening of the cell sap indicates that the acid penetrates immediately. As the more or less complete return of the original hue to the cell sap, after rinsing with water, indicates, the acid can be diosmotically removed from the cell again.

If, in beet sections, the cell contents were first contracted with 20% sucrose solution and then the sucrose solution under the cover glass was replaced by a solution of the same concentration (but containing HCl in the ratio given above), then normally 2-5 hours later the protoplasm's ability to expand had already been partially or totally abolished.[19] When more dilute sucrose solution was added, the pigment of the cell sap was quickly lost, for reasons already stated. On the other hand, when the concentration of the surrounding sucrose solution is carefully kept constant, cells seem not to give up any of their pigment for days. In particular, when a sucrose solution that contains hydrochloric acid is replaced after a few hours of action by a sucrose solution of equal concentration, it is possible to find cells with cell sap whose pigment has not changed some five or six days later. In the end, of course, all the cells have lost their pigment, but this loss from the cells occurs not, as one might suppose, quite gradually, but always quite quickly. This is no doubt only after the plasma membrane has been torn or undergone a change that modified its diosmotic properties. After all, it is to be expected that once life has been abolished the (what is most likely) albuminous material will finally decompose. Incidentally, treatment with water containing hydrochloric acid produces the same result. After being acted upon for several hours, the cells can no longer be contracted by sucrose solution and the pigment escapes from them.

Analogous results were obtained, by treatment with hydrochloric acid of the specified dilution, in petal cells of Pulmonaria officinalis and Anchusa officinalis (in 14% sucrose solution), and with vacuoles of the protoplasm of Vaucheria, Aethalium and the root hairs of Hydrocharis (in 2-6% sucrose solution). The vacuoles were formed and observed in sucrose solution stained with aniline blue or with cochineal. The result with red beet and with vacuoles of Vaucheria protoplasm was similar when acetic acid or sulfuric acid, of roughly the same dilution, was used instead of hydrochloric acid.

If the protoplasm forms sufficiently thick layers and if it is suitable for observation, as it is in younger root hairs of Hydrocharis and in larger masses of protoplasm that have flowed out of Vaucheria, then it is possible to observe the visible changes in the protoplasm cuased by the hydrochloric acid that has penetrated. These changes take place in the same way in samples immersed in sucrose solution and in samples containing non-contracted cell contents. The streaming of the protoplasm of Hydrocharis hairs is stopped shortly after adding water that contains hydrochloric acid. Soon thereafter, the protoplasm is often seen gradually becoming cloudy, and very soon, or no more than an hour later, the protoplasm has been transformed into the kind of cloudy, granulated mass we are familiar with for protoplasm that has been killed. In spite of this, dyes such as aniline blue, cochineal and hematoxylin penetrate it no more than they previously penetrated the living protoplasm. It has of course been reported that pigmented cell sap is retained in HCl treated cells. The protoplasm of such cells remains colorless and also shows the granulated state. Obviously, the surrounding liquid must be kept absolutely at the same concentration, for when the plasma membrane bursts dyes and pigment can enter and leave freely. Now the protoplasm, which up to that point had been colorless, accumulates dye in the same manner as dead protoplasm.

The facts I have just discussed make it clear that the surrounding periperhal, what appears to be membrane-like, covering of protoplasm that is dead in appearance and behavior keeps dye from entering. I was able to observe this directly in a very conspicuous manner. When, by diluting the surrounding solution, the osmotic pressure is raised in a contracted protoplasm body, whose ability to expand has been destroyed, the plasma membrane sometimes tears only in one place. Accordingly, the pigmented cell sap--say, from the cell of a red beet--is seen emerging at one point and spreading in the surrounding sucrose solution. It is less easy to observe the penetration of pigment into the protoplasm (of beet cells). However, I succeeded several times in observing this process in young root hairs of Hydrocharis whose protoplasm body had been contracted with sucrose solution and deprived, by HCl, of its ability to expand. Here the pigment not only penetrated the cell sap from the tear produced by osmotic pressure, but it also gradually spread from that tear into the dead protoplasm enclosed between the plasma membranes. In one case the dead protoplasm was already almost stained before pigment entered the cell sap, apparently because a tear had first formed only in the outer plasma membrane. Of course here, as generally in most cases, the tear was not directly visible, but after what I have said it is not necessary to give particular proof of its existence. It is quite obvious that such experiments can produce favorable results in isolated cases only, even if one is very careful.

For the present, the observations I have just shared with you offer exact proof that the peripheral, what appears to be membrane-like, layer that covers the dead (coagulated) protoplasm shows diosmotic properties, as does living protoplasm, while in the enclosed coagulated protoplasm pigment spreads easily. The fact that this peripheral layer tears when osmotic pressure is increased shows that it is unquestionably a membrane, i.e. that it possesses the cohesion of a solid. Since this tearing can occur before the protoplasm body has grown noticeably in size, it is apparent that this plasma membrane is usually able to expand only very slightly. For true increase in this membrane area is obviously caused by the fact that the gradual destruction of the membrane-forming material has not yet been completed and that therefore limited growth of the membrane is possible. From my observations I cannot draw more detailed inferences about the ability to expand and the elasticity of this plasma membrane, yet the above is enough to exlude the possibility of an aggregate state that is other than solid.

Living protoplasm is inconceivable, if lacking the formation material for plasma membrane. Since it has until now been impossible to remove this material or to render it innocuous without destroying the living state, and perhaps it is impossible anyhow, the cohesion of the peripheral layer could not be determined as long as it is connected with viable protoplasm. It was thus impossible to decide by such an approach whether the plasma membrane, thought of as isolated, is unquestionably a solid membrane. If we now consider the diosmotic properties of the living protoplasm body, there can be no doubt, to begin with, that even the peripheral layer does not allow those substances to pass through which are incapable of penetrating living protoplasm. This can be seen directly with dyes; for other substances it has to be inferred by necessity. For when a solute enters the inner streaming parts of a protoplasm body, it must also be distributed in it as much, or perhaps far more, than solid bodies that are mechanically tossed about. Obviously, we should disregard cases where a new factor for distribution is added, for example, by binding to certain parts of the protoplasm. Besides, in living cells, no matter what the aggregate state and structure of the protoplasm, this protoplasm is not a really solid body; rather, it has a high water content and has comparatively little solid substance. For such a body it is actually unthinkable that an easily diosmosing and otherwise neutral solute should not spread. In very slimy and gelatinous bodies, as we know from experience, crystalloids diffuse and diosmose with the greatest ease.[20] In order to prevent the diosmosis of crystalloids, all the water-filled spaces between the constituent particles of a body need to be correspondingly narrow (see p.137).

The diosmotic properties of living protoplasm,[21] which have been observed in fact, are actually only such properties as are also exhibited by certain of Traube's precipitation membranes. Now that we know of the latter, those properties can no longer be regarded as exclusive characteristics of the living organism. Right from the start, what came to mind was that in the peripheral layer, which in fact decides whether a substance is absorbed or not, there might be a membrane which, like a precipitation membrane, might even have minimal thickness, obviously without impairment of its qualitative diosmotic behavior. The circumstance that has already been mentioned--that crystalloid substances, which are unable to penetrate the peripheral layer of the protoplasm, diosmose easily through slimy and gelatinous bodies--speaks in favor of a membrane. Moreover, it becomes at least probable that this diosmotically decisive layer has the solid aggregate state, seen from the following standpoints. The peripheral layer is formed from a substance that is solid in its anhydrous state. The interstices, which are necessarily narrow because of the diosmotic properties, demand that the particles (tagmas) that constitute the peripheral layer be densely packed. Also, protoplasm that has been killed, as described earlier, is covered by an unquestionable membrane, whose diosmotic properties are identical with those of the peripheral covering of the living protoplasm. On the basis of all these considerations, it is highly likely that even the living protoplasm body is already surrounded on all sides toward the cell wall and the cell sap by a true membrane, which is resistant when it is not capable of growth. All that is essential has already been said about its thickness, which may perhaps be only minimal, and its indefinite delineation against the hyaloplasm.

In evaluating diosmotic processes in living cells, it is, incidentally, a matter of indifference whether the peripheral covering of the protoplasm is regarded as a resistant membrane or as a viscous layer of slime. At any rate, because of its ability to grow, this covering reacts to a mechanical expansion like a viscous substance. When facts are soberly examined, there can be no doubt that the familiar diosmotic properties of protoplasm are determined by the peripheral layer. A similar appearance, a similar formation, identical diosmotic and other properties admit hardly any doubt that the hyaline membrane-like layers around living protoplasm bodies and non-living vacuoles, such as those formed by protoplasm when immersed in water, are equivalent in their physical structure.[22] Other diosmotic properties, which I shall discuss below, also speak in favor of such equivalence, even for the type of plasma membrane that surrounds the coagulated protoplasm incapable of promoting growth. Of course, a number of consistent diosmotic observations cannot make the physical and chemical identity of two membranes into an unquestionable certainty. It is of course possible in our case that the very same causes which produce decompositions in the enclosed protoplasm also affect the membrane, without however causing such structural changes as those which produce a conspicuous change in the diosmotic properties. We lack actual data to decide this type of question definitively.

Reaction to dilute acids and alkalis, and to mercuric chloride and iodine, was found to be completely identical both for the plasma membrane that encloses living protoplasm and that which encloses dead protoplasm. As stated on p.148, change in color in pigmented cell sap indicates the rapid diffusion of even extremely dilute hydrochloric acid, the same being true for acetic acid and sulfuric acid. Furthermore, the acid can again be disomotically removed and the previous color restored. The same is also true of ammonia, to which I exposed plant cells, having diluted it greatly (one drop of ammonium hydroxide per 15-30 cc liquid). Reddish cell sap in Pulmonaria petals and filament hairs of Tradescantia immediately turns blue,[23] and as more ammonia is added it gradually takes on a greenish and grayish color.[24] Yet here too the blue color returns slowly, after rinsing with water, if the ammonia has not been acting too long. Besides, the protoplasm was not killed during this diosmotic exchange, because the protoplasm streaming in the filament hairs of Tradescantia, which the action of the ammonia had stopped, is gradually restored to normal levels again as the ammonia is removed.[25] Dilute solutions of potassium hydroxide and potassium carbonate have a very similar effect.

If dilute acids and alkalis do not alter the diosmotic properties of the plasma membrane, this does occur in a very striking way with iodine and mercuric chloride. Very dilute solutions of each (I used a mercuric chloride solution of 0.2%-0.5% concentration, and water slightly tinted by iodine) quickly penetrate the protoplasm and cell sap, as indicated by the very distinct color change that mercuric chloride produces in pigmented cell sap, and by the staining of proteins and, perhaps, of grains of starch by iodine. Soon a very gradual diosmotic loss of the pigment begins, and this proves that what is involved are only molecular changes in the plasma membrane and that the membrane has not been torn. The time required for the loss of pigment may extend to hours if the above-named reagents are rinsed out after reacting for a brief time. It is also clear in this case that the pigment does not disappear because it is destroyed, but by diosmosis. At this time, I am unable to decide whether iodine and mercuric chloride already diosmose through the normal plasma membrane, or only pass through after causing molecular changes in it.[26]

From the facts I have already discussed, we can draw the very general conclusion that substances contained in the protoplasm and cell sap do not significantly diosmose, even through the membrane that covers coagulated protoplasm. For a decrease or increase of osmotically effective substances within the protoplasm or cell sap should also bring about an altered osmotic pressure, in the negative or positive sense, thus bursting the plasma membrane, which is no longer able to grow and is not pressed against a support. As we know from experience, this does not happen. In actual fact, however, osmotic overpressure produced by a slight difference in concentration suffices to cause the membrane to burst once the ability to grow is totally destroyed. The plasma membrane of the red beet cells readily burst when the sucrose solution used for contraction was replaced by another solution containing 0.25% less sucrose. What I just said must be kept in mind when evaluating experiments where the conditions required for the plasma membrane to grow were abolished. For through coagulation of proteins, and also through other chemical processes, the osmotic pressure of the cell content will vary and can cause the plasma membrane to burst if these fluctuations in pressure take place after the plasma membrane is no longer capable of growing. In this respect, of course, specific differences are possible in dissimilar cells, and it is conceivable that such pressure conditions were a contributing factor in the final loss of color from the cells I used in my experiments.

The plasma membrane, once formed, is strikingly resistant toward reagents. The plasma membrane remains undissolved, although sometimes in fragments, after extended exposure to dilute acids and alkalis at normal temperature and with boiling. At least I found this to be true for filament hairs of Tradescantia, root hairs of Hydrocharis and also for the vacuoles just formed out of Vaucheria protoplasm. Incidentally, in all these cases, some parts of the remaining protoplasm body were still undissolved as well, as could be seen especially after staining with iodine. In spite of this resistance to the above-named and other reagents, the protoplasm is able to dissolve the plasma membrane, or at least to transform it into a softened state, which allows a distribution of the membrane substance in the protoplasm. This follows from the coalescing of plasmodia, zoospores and other cells. As a matter of fact, I was unable to find a trace of membrane fragments, either directly or after treatment with potassium hydroxide and acid, in the [*protoplasm*] mass that had been formed by the coalescing of minute plasmodia of Aethalium.[27] Incidentally, I was also able to follow the coalescing of the very thin plasma layer of two vacuoles formed from protoplasm of the root hairs of Hydrocharis, in 3% sucrose solution. It follows that the dissolution of the plasma membrane does not directly depend on the activities of life in the organism but, as could have been expected, is dependent on the existence of a specific solvent medium.

However, if the protoplasm is able to dissolve the substance of the membrane,[28] the latter will necessarily be able to attain only a certain thickness, which results from the competition between the dissolution from within and the new formation by precipitation after contact with water (see p.139). It is essential that we take this process into consideration when interpreting the familiar phenomena that are observed when protoplasm bodies are contracted by dehydrating agents. As we know, contracted protoplasm bodies do not have a wrinkled surface even if the volume decrease has been considerable. This circumstance would be convincing evidence against the existence of a resistant membrane that is only slightly elastically stretched, except that the same phenomenon, because of the dissolving action of the protoplasm, could be produced with a thin precipitation membrane.[29]

We must agree that a precipitation membrane has a minimal elastic tension, even when membrane-formers are present. As the force of tension then lessens, it is therefore inevitable that the membrane thickens very slightly. In our case the thickening of the membrane is balanced by the dissolving or loosening effect of the protoplasm which, all other things being equal, must reduce the membrane to its original thickness. As individual membrane particles (tagmas) are partially or totally dissolved, there is a change in the balance among the molecular forces that determine the relative positioning of the tagmas. With the transition into a new state of balance, a shifting of the tagmas is natural even in a solid membrane. If these processes are continuous, they may produce unrestricted decrease in the membrane surface while the membrane retains the same, minimal thickness.

I must leave open the question whether the irregular

surface shape that protoplasm bodies often exhibit during contraction is related to the dissolution process we have been discussing. It is just as likely that there are other reasons for this phenomenon. Be that as it may, if we admit a membrane that is resistant in itself, the relatively rapid rounding of the surface demands that the process whose underlying principle we have just developed should take place relatively quickly. Still, I have no objection to this assumption since, after all, the removal of constituent tagmas may take place just as quickly as the opposite process, the insertion of new tagmas when, as may sometimes happen, the membrane area grows very rapidly.[30] I shall pass over a number of questions that are related to the topic discussed here, since I was basically interested only in demonstrating that contraction phenomena in the protoplasm simply cannot be used as an argument against the presence of a membrane that is resistant in itself.

Though I cannot pronounce a definitive judgement on the chemical composition of the plasma membrane, it is probable that, as Mohl (1844c and 1855:694) had already assumed for the primordial utricle, proteins play at least a partial role in the structure of the plasma membrane, if the latter does not consist entirely of proteins. Though it is hard to judge with thin objects, I believe I can perceive a yellowish-brown staining with iodine in the entire hyaline border layer, where the latter is so thin it may likely be nothing but plasma membrane. Iodine in zinc chloride gives the same reaction, both when used directly and when the material has first been treated with potassium hydroxide and acid. Aniline blue also seems to be accumulated. On the other hand, I cannot say whether with Millon's reagent and with nitric acid the usual reaction for albuminous substances occurs.

Like the reactions just described, the reaction of the plasma membrane to mercuric chloride described previously (see p. 153), which definitely indicates a chemical combination of mercury with a constituent part of the membrane, suggests that albuminous substances are present. At least it is a fact that the latter combine chemically with mercuric chloride, while carbohydrates and pectins, which might be assumed to be other important constituent parts of the membrane, are not known to do this.

The fact that the plasma membrane is insoluble in moderately dilute potassium hydroxide does not, it is true, agree with the behavior of those proteins that have until now been produced form plant organisms. Yet seemingly albuminous substances, which behave similarly, such as chitin, elastin and the like, have long been known in the animal kingdom. These substances cannot be obtained at all by the usual preparation methods. In actual fact, protoplasm very commonly, though not always, seems to leave a residue of insoluble substances, in addition to the plasma membrane, when treated with moderately concentrated alkalis and acids. These insoluble substances exhibit normal reactions for proteins.[31] But until these substances have been chemically analyzed, one can draw no definite conclusions on the basis of the reaction to the reagents mentioned. Thus, for example, we cannot rule out the presence of pectins only because all substances know until now which belong to this chemically still obscure group are transformed into a soluble form by potassium hydroxide.

As long as the chemical composition of the plasma membrane and the membrane-former is not definitely known, it will hardly be possible to discover all the facts about the dissolving medium present in protoplasm. The reaction of the plasma membrane with potassium hydroxide and other reagents can, of course, raise no doubts about the existence of a solution of membrane-former and the dissolving action of protoplasm on the plasma membrane.[32] We must leave open the question as to the agents by which that dissolving action comes about. If proteins are involved our first thought will be pepsin or similarly acting substances, since pepsin actually dissolves coagulated proteins rather quickly in certain conditions, while these are hardly dissolved by dilute potassium hydroxide.

The abolition of conditions that enable the plasma membrane to form is probably caused by the chemical decomposition of the solution of membrane-formers, without diosmotic loss of the dissolving or the dissolved substance, when the plasma membrane growth is abolished while there is no change in the diosmotic properties. An argument in support of this view is the rapid effect of hydrochloric acid, as compared to the relatively long time it takes to reach the state under discussion without the help of reagents. I assume that the membrane-former, which was previously dissolved, is among the substances precipitated by hydrochloric acid in the protoplasm. The presence of a small amount of potassium hydroxide or ammonia with sodium chloride, in the surrounding medium, does not abolish plasma membrane growth. Nor is its formation prevented by their presence, as we see when protoplasm masses are crushed in a solution with contains those substances in solution. I was led to perform these experiments by the fact that in egg white a small amount of protein is kept dissolved by salts, and is more or less completely precipitated both after dilution with water and after dissolving salts are diosmotically removed.[33]

According to empirical experiences, a plasma membrane will exist wherever protoplasm adjoins another aqueous liquid but, as I showed earlier, we cannot say with certainty that plasma membrane is formed only under that condition. We must also keep this in mind when we ask the question whether structures that occur within the protoplasm body and that themselves consist of protoplasmatic material (such as the cell nucleus and the pigments) are bounded by a plasma membrane. As the following shows, this question is important in a number of ways.

When an isolated cell nucleus or chloroplast is placed in a sucrose solution capable of just noticeably separating the protoplasm body from the cell wall, its form and appearance are essentially the same as observed for these bodies in the protoplasm. When the sucrose solution is diluted, the volume of these structures increases and, in pure water, disorganization is the final result,[34] as we know. The presence of plasma membrane is inferred from the above mentioned observations on diosmotic behavior and from the osmotic effect of the contents, which we just described. At the same time, the increase in surface area thus brought about shows that there is usable formation material available for the plasma membrane. Obviously, it is impossible to ascertain from these observations whether the plasma membrane already exists within the protoplasm, and at present I cannot answer this question. It is obvious that we face special difficulties here; particularly, no conclusion as regards the plasma membrane can be drawn from the fact that dyes that in themselves are perhaps soluble in water are limited to certain pigments.

The existence of a plasma membrane would not be possible if, in the nucleus (or chloroplast) and the surrounding protoplasm, formation material and the solvent for the plasma membrane would be present in the exactly identical way. For this very reason the question I have raised deserves very special attention. To be sure, these differentiated structures are certainly not qualitatively identical and during differentiation itself there might be cause for delimitation by a special peripheral covering. Be that as it may, there is a series of unanswered questions here that I do not intend to dwell on. In fact, as I stated earlier, quite a few obscure points still need to be settled as regards the causal origin of the plasma membrane (or the hyaloplasm). That is why I should like to express an emphatic warning not to try to give premature causal explanations for the appearance of membrane-like layers during cell formation and cell division. On the basis of our experience we can only state that the plasma membrane (and perhaps hyaloplasm as well) has to form in circumstances where a separation is produced by some unknown vital event and, as a result of which, masses of protoplasm are embedded in another aqueous medium that does not have a dissolving effect on the plasma membrane. We cannot say, however, that a separation of this type is necessary for a plasma membrane to be produced.

When membrane-like layers are found around bodies present in cell sap, these layers, according to our earlier definition, must not be described as plasma membrane, even if they should be identical with the plasma membrane chemically and physically. This would be probable when colorless vacuoles float in cell sap (Nägeli, 1855:9). An unquestionable precipitation membrane, whose mode of formation remains to be studied, covers the drops of tannic acid that are found, especially beautifully, in the articular cells of Mimosa pudica.[35] I must leave open the question whether the oil bodies of liverworts, grains of protein and other structures in the cell sap of the living cell are covered with a diosmotically decisive membrane. Understandably, conditions for the formation of a precipitation membrane do not exist in the aqueous cell sap in the same way as they exist in protoplasm. Yet, it must still be decided whether this is the case when the cell sap becomes richer in proteins, as in maturing seeds. If the answer is yes, we would have to ask the same type of questions as those about the delimitation of structures found within protoplasm.

Necessarily, one question will suggest itself. Are the precipitation membranes found in and around protoplasm, and also in cell sap, physically and chemically identical? We have no evidence to judge the chemical composition, and the diosmotic properties observed until now have not shown a physical difference. But this has little significance since observations had to allow small differences to be overlooked and because, moreover, the diosmotic lack of equivalence can, to be sure, be demonstrated by a single positive result, but cannot be disproved beyond doubt by a few negative results. Indeed, diosmotic differences must be regarded as probable rather than improbable, if only because our membranes adjoin media of unequal composition, in several cases. These might possibly modify the properties by causing the membrane to swell or shrink, or by causing a membrane to form whose molecular structure is different. It is possible that some substance, which is not present in every case, combines chemically with the membrane-former. Further, we should remember that infiltration, i.e. the insertion of foreign particles, can change the diosmotic behavior of precipitation membranes, as Traube (1867:141) showed.

19. Comments on Molecular Structure.

If we regard the plasma membrane as a precipitation membrane, we will also recognize that it has an analogous structure, i.e., we will consider the plasma membrane as a syntagma. This is quite likely since proteins, of which the plasma membrane seems to consist, are surely colloids (Graham, 1862:61). We could also put forward the argument that up to now precipitation membranes are known only for colloids in which, as we discussed earlier, we must assume that molecules have been layered together into tagmas (see p. 34). At present, it is not possible to say with any degree of certainty whether the constituent tagmas contain water, or whether the hydrated precipitation membrane contains only intertagmatic water. Also, the structure of the tagmas must be left undecided; after all, their chemical structure also needs to be determined conclusively.

An artificial precipitation membrane is also an "organized body" whose molecular structure, according to Nägeli's conjecture, is only a special case of syntagma. Syntagma is a term we used for a substance composed of tagmas, whether it is capable of swelling or not (see p. 35); however, when a syntagma is capable of swelling in a limited manner, we have an organized body in the sense Nägeli intended.

Nägeli included a crystalline, or at least polyhedric, form of tagmas in his definition of organized body as he later formulated it.[36] But this contradicts his own perceptive ideas, according to which tagmas in young starch grains and cell walls [*Zellmembranen*] must have a spherical form (Nägeli, 1858:336, 361). It is a fact that a limited ability to swell and other observable properties of organized bodies are compatible with every form of tagma. The requirement that water should be absorbed only intertagmatically, but not into the tagmus themselves, also must be abandoned. There is no question that water can enter the constitution of a tagma (see p. 36), and Nägeli (1858:353) himself formerly mentioned this possibility, but later no longer discussed it. If we admit that there is bound water, bodies will lose this water, or part of it, when they are dried and reabsorb it when water is again added. This happens in fact with many hydrates of colloidal bodies. When such a substance again absorbs water and returns to its original state there is no reason for not calling it organized, if it is so according to its other properties. Obviously, where water content fluctuates in this way, the constitution of the tagma changes temporarily. Thus, the definition of an organized body should not be formulated so as to include an absolute constancy of composition under variable conditions. Nägeli himself considered the structure of organized bodies, formed from two chemically different substances, only as a peculiar characteristic that accords with present experience. Therefore it need not be given special consideration here.

The cohesion of a syntagma penetrated by a liquid is a function of the mutual attraction of the tagmas among themselves and of the tagmas' attraction for the liquid. Because of these variables, the cohesion may change and in a specific medium the tagmas will be able to distribute in such a way that a solution is formed, while the same body [*syntagma*] in another liquid swells only in a limited way. Since this is obvious there is no need to demonstrate this with specific examples, but gum arabic or dextrin, with water or with alcohol that contains water, might provide a case in point. In other words, a body can be organized in relation to one liquid, while it is not organized in relation to another liquid, and we must keep this in mind when forming an opinion about the structure of bodies found within the organism, which are mostly immersed in solutions, not in pure water.

Thus an "organized body" is a syntagma in relation to a specific liquid. It is capable of limited swelling, which can be reversed, and in the process it absorbs liquid (at least into the intertagmatic spaces but possibly also into the tagmas themselves) or gives off water, respectively, from the tagmas or the intertagmatic spaces. Incidentally, except for some extensions and limitations, this is basically Nägeli's definition of organized substance. Organized substance, as Nägeli believes, is always characterized only by a specific molecular structure, but not by the location where it occurs, and there is no a priori requirement that all formed organic constituent parts of an organism be organized. Brücke (1861:386), on the other hand, associates no very specific physical idea with the term "organization". On the contrary, that term is merely intended to indicate that there is an unknown structure in the living organism that is specific to this organism. In this sense the living organism (or its parts) also has often been described as an organized substance. I refrain here from a discussion on whether it is not more convenient to use this latter interpretation of the term "organization," and merely note that, for the time being, I will use "organized body" in Nägeli's sense, and that the term "syntagma which is capable of swelling" will be used as its synonym.

Is protoplasm organized or not? To answer that question we need to consider the surrounding medium and, of course, other additional factors that are related to the existence of the enclosing plasma membrane. As a consequence of the plasma membrane, osmotically effective substances produce a pressure that causes many protoplasm bodies exposed to water to expand. This brings about the final destruction of the molecular structure, although none of this happens if the surrounding liquid has the appropriate concentration. The plasma membrane also prevents the loss of solutes from the protoplasm, which would be followed by a far-reaching structural change of the protoplasm. To rephrase the question--is protoplasm an organized substance under the conditions prevailing in the viable cell? Naturally, this purely physical question at first needs to envisage only a static state of the type that may actually exist in non-viable or inviable protoplasm. For continuous change, which is associated with the activities of life, is always only a consequence of constant disturbance of the state of equilibrium by a still obscure interplay of forces.

Sachs (1865:443) is probably the first to have tried to apply Nägeli's view of the molecular structure of organized substances to the protoplasm or the ground substance of the protoplasm body, respectively. Those arguments that are based on a sharp demarcation between the protoplasm and other media, and on diosmotic properties, are now untenable, as I need not specifically prove, and the cohesion of the protoplasm cannot be used to decide whether tagmas are present or not. The circumstance that the ground substance of the protoplasm apparently consists of proteins makes it likely that tagmas exist. Because the proteins are colloidal substances, we shall assume that their molecules are aggregated into tagmas.[37] The fact that vacuoles are formed when a large amount of water is added indicates that the protoplasm body, which is covered by plasma membrane, can only absorb a limited amount of water. Incidentally, this is not an irrefutable argument, though I will not specifically show it here. Under the conditions existing in the cell, the uptake of water into the protoplasm has certain limits imposed on it by osmotic effects, as I shall explain in the following. Accordingly, the protoplasm body, or the ground substance of it, will be called an organized body because it is composed of tagmas and has a limited capacity to swell under the given conditions (the term "organized substance" will be used for this unless a truly solid aggregate state is not required). For the protoplasm, which has a high water content, is not absolutely solid and the resistance it shows to expanding forces resembles, at the most, that of a somewhat gelatinous substance. We should not rely on the fact that some of the formed protoplasmatic bodies have a higher cohesion, for the term "organized" refers not to the chemical properties of the material, but to its physical structure. Furthermore, the same chemical substance might be organized or not organized depending on the cohesion of its tagmas. Incidentally, I myself would like to call a body, even if it is not really solid, "organized" if it only has a tagmatic structure and (under given conditions) takes up a limited amount of water, and thus of protoplasm as well.

Where organization is concerned, the situation remains basically the same when only some part of the protoplasm, whether ground substance,[38] or hyaloplasm, or some other delimited strucutre, is to be considered. Visually perceptible structural relationships, of the type sometimes observed in the hyaloplasm,[39] can only speak in favor of organization, and probably also for polyhedric tagmas. When thinking about these questions, one thing has sometimes not always been considered. A visible structure is also possible when molecules, not tagmas, are the ultimate constituent parts; i.e., the type of organization cannot be inferred from structure alone.

I felt I needed to make the above remarks following what I said about the plasma membrane, but I think I need not prolong the discussion. Because of insufficient data, I also do not want to go into the question of what closer relations exist among the plasma membrane, hyaloplasm and ground substance of the protoplasm.[40] The potential ground substance of the protoplasm cannot be absolutely identical (i.e. identical physically as well) with the plasma membrane as regards diosmotic processes. If we then keep in mind that the building material for the plasma membrane might be contained as a solution in the protoplasm, we already have a series of points of view which, combined with other considerations, can probably provide us with means and starting points for explaining the questions just raised.

20. Diosmosis through the Plasma Membrane.

The studies discussed in the physical section were done in order to obtain means for assessing the osmotic processes in the living cell, whose protoplasm body, as has been shown, is covered with a plasma membrane,[41] which has diosmotic properties similar to certain artificial precipitation membranes. Now that this fundamental preliminary work has been done, future research will have the task of studying osmotic processes in the organism to reduce the phenomena, depending on these processes, to their causal conditions. Many, and often very difficult, questions need to be solved here, questions that touch on the most important problems in physiology. After all, osmosis plays a prominent part in nutrition, growth and other processes.

The sections that follow propose not so much to present new empirical studies as to show to what extent well-known facts admit of an explanation, and also to throw light on questions that are as yet unresolved. I hope thus to provide an impulse for others to become active in a field which is so vast it is too much for one person's energy.

Let us first look at the diosmotic exchange of substances, focusing on a cell covered by a cell wall in which the protoplasm forms a simple wall layer. Obviously, a solute can enter the protoplasm only if it is able to diosmose through the cell wall and the adjacent plasma membrane. In order to enter the cell sap this substance must be distributed in the protoplasm and also be able to pass through the plasma membrane which delimits the protoplasm against the cell sap. This delimitation is always fully complete, whether the protoplasm forms a simple wall layer, or permeates the cell space in threads or bands, or whether the cell sap is distributed in numerous vacuoles. The situation basically remains the same no matter what form the protoplasm body has. The diosmotic properties of a hydrated cell wall, as experience shows, are such that the wall certainly lets through all those substances that are able to pass through the plasma membrane. Thus, with respect to the diosmotic uptake into the protoplasm, the same is basically true for a cell covered with a cell wall and for a primordial cell. Incidentally, we must exclude cuticularized and suberized cell walls and, generally, such cell walls as hardly, or very minimally, imbibe water. For these will prevent substances that diosmose through the plasma membrane from passing through. Proof that the plasma membrane determines the diosmotic properties of protoplasm was presented earlier. Reasons were also given why a solute, if it diosmoses through the plasma membrane, must also spread in the protoplasm unless chemical binding inhibits its spreading. As I showed, this follows from the visible phenomena of movement in the flowing protoplasm, in which even unreactive solids are tossed about. It also follows from the circumstance that the aggregate state of protoplasm can at most somewhat delay the diosmotic diffusion of crystalloids (and only crystalloids are able to pass through the plasma membrane), but cannot abolish it. These considerations leave no doubt that solutes which diosmose through the plasma membrane must necessarily also spread in the protoplasm. I do not know of any soluble pigment substances that are constituent parts of the protoplasm and could demonstrate such diffusion directly, by the way they spread. Perhaps, however, attempts to place granules of soluble dyes in larger protoplasm bodies might succeed without damaging the latter--after all, an injury is immediately closed again by the plasma membrane (I did not do this experiment). The way the dye is distributed might also give other information on the structure of the living protoplasm body, because only undissolved proteins are able to accumulate dyes. We cannot draw a compelling conclusion for the question we are considering here from the fact that substances that penetrate the plasma membrane, such as ammonia and hydrochloric acid, also rapidly spread in the protoplasm. Diffusion would take place even if the diosmotic properties of the plasma membrane and protoplasm were simply identical.

Perhaps it is not quite superfluous to emphasize once more that plasma membranes that enclose dead and living protoplasm have identical diosmotic properties, judging by observations that have been made, and that this identity also exists between living and dead protoplasm. For, both living protoplasm bodies and vacuoles formed from protoplasm fragments show the identical diosmotic behavior in every conceivable state of expansion caused by osmotic pressure. Similarly, the diosmotic properties remain unchanged if the activities of life are terminated in some way, e.g., by exclusion of oxygen. When we are dealing simply with uptake or lack of uptake of a substance into the protoplasm, the specific structure of the plasma membrane and activities of life need be considered only insofar as the physical structure of the plasma membrane might be modified by it.

In strict consequence of the above considerations, protoplasm must take up any substance that diosmoses through the plasma membrane and, furthermore, any substance that is not retained in the protoplasm by binding must also enter the cell sap, if the plasma membrane that surrounds the cell sap allows diosmosis. This uptake (or discharge) must continue for as long as the difference of concentration within and outside the cell causes unilateral diosmotic movement. The limit of uptake is thus determined both for an indifferent substance and for a substance that takes an insoluble form in the cell, or which goes through any kind of transformation at all. For such a substance, uptake must continue as long as reaching an osmotic state of equilibrium is prevented.

The kind of exhaustive experimental proof needed to satisfy the above, strictly logical, demands has not been obtained. Still, I would like to point out that we know from experiments that a number of inorganic substances not needed by the plant are taken up by it. For the alkalis that form no insoluble salts, such as cesium, rubidium and lithium, we may assume with certainty that they are found not only in the cell wall, but inside the cell as well. Also, the diffusion within the cell of certain substances that diosmose through the plasma membrane is established, e.g., for hydrochloric acid and ammonia. Similarly, as for ammonia, the uptake of lithium carbonate, which is not necessary to the plant, can be observed when red cell sap turns blue. Of course these, like iodine and mercuric chloride, are not neutral substances. Yet, very dilute solutions of alkalis do not destroy life. Protoplasm streaming stops only some time after the uptake of such a substance, and when it is diosmotically removed the streaming returns very soon.

Except for dyes and substances which reveal their uptake by a reaction in the colored cell sap, very little is known about the diosmotic properties of the plasma membrane. In order to observe the passage of small quantities of other substances directly, the methods we have used so far are not adequate. During a longer period of time, however, even a minimal diosmosis will be able to move larger amounts of a solute into a cell, all the more so because the area of the plasma membrane is relatively large when compared to the small volume of the cell. But, diosmotic uptake into living cells involves the entire area of the plasma membrane, provided that it is in contact with any solution that saturates the cell wall. For, of course, it is certain that there will be uptake of inorganic and organic substances into the cell, but it is obviously impossible to infer in what form the substances diosmosed from the fact of their presence.

In order to test the permeability to certain salts of the protoplasm body in the cells of red beets, deVries (1871a) contracted the cells with solutions of the appropriate concentration and observed whether the state of contraction of the protoplasm body remained unchanged. If it did, either no diosmotic exchange had taken place at all, or the exchange between the salt solutions and the substances contained in the cell must have been regulated in such a way that the osmotic effect remained unchanged on both sides. For, in all other cases, the state of contraction of the protoplasm would obviously have to change and if the volume of the protoplasm increased, this would always indicate uptake of salt into the cell. If the uptake of the salt was considerable, the protoplasm body would eventually once more be lying aginst the cell wall [*Zellwand*]; however, deVries did not see this happening when he used different magnesium, potassium and sodium salts. Since the cell contents of living red beet cells give up only slight (if any) amounts of water, deVries' observations make one thing certain; minimal amounts, at most, of the salts that were used penetrated the cells diosmotically. To decide this question, very accurate measurements would have to be carried out; deVries did not do so. Because of various technical difficulties, it would acutally be impossible to determine this very minimal uptake of a substance into the cell by means of such measurements. In principle, the method is quite correct. It is practicable when there is sufficient disomotic exchange and when solutions of colloids are used for contraction--i.e., substances which surely do not pass through the plasma membrane--it could also be decided whether the protoplasm body diosmotically releases the substances contained in it to the outside.

In order to test accurately whether sucrose actually does not diosmose out of red beet (Hofmeister, 1867:4), a cylindrical piece about 10 cm^2 in surface area was cut out of the inside of a red beet containing a large amount of sucrose. After the piece had been most carefully washed for an hour by repeatedly renewing the water around it, it was left for six hours in 100 cc water. The volume was reduced by boiling to four cc, a little hydrochloric acid with Fehling's solution was added, and it was brought to a boil. There was no trace of sucrose reaction. Therefore I am sure that, in the form in which it is contained in the cell,[42] sucrose can, at the most, diosmose through the plasma membrane in extremely minute quantities.

With such a negative result as obtained for sucrose, it is out of the question that cells are killed during the duration of the experiment. If the facts are properly evaluated and analytical methods are accurate, such a check can serve to prove observed diosmotic release of a substance beyond doubt in cases where positive results would necessarily be questioned because there is a possibility that cells have been damaged. When experiments are properly conducted and a sample is first immersed in a solution of the substance, then rinsed and placed in pure water, it is possible thus to decide whether there has or has not been uptake. Where tests permit, it may be possible to ascertain only a slight degree of diosmosis.

Incidentally, other microscopic methods of deciding whether there has been uptake of a substance are also available. Generally, they can be used when a solute shows directly, by a visible reaction that it has penetrated a cell, or when its presence in the living cell can be determined by the subsequent addition of another substance. For example, when cell sap that reacts as an acid is made alkaline by ammonia, which is possible without killing the cell, a precipitate must necessarily form if substances are present that are soluble in acid but not in an ammonia solution. This remark should suffice. I merely wanted to indicate the principle of methods that, if they can be carried out and if material is skillfully combined and properly selected, may give specific information on the distribution of solutes within the cell and on the chemical composition of protoplasm and cell sap.

The diosmotic properties, obviously, are dependent on the state of the plasma membrane at the time, and it is not possible to say a priori that its state is always the same. On p. 160, I pointed out the possibility that infiltrations or contact with media of a different composition could, by shrinking or swelling,[43] have a modifying influence. These questions may be of importance for the uptake or migration of substances in specific cases, but we know just as little about them as about the influence of temperature on the diosmotic properties of the membrane. We must also leave undecided the questions whether plasma membranes of different cells always have identical properties, and whether the interior and exterior side of the same membrane are equivalent.[44]

The thickness of the plasma membrane, which may not be the same in all cases, probably does not affect diosmotic exchange qualitatively, but will affect it quantitatively, and thus may possibly need to be taken into account as regards the uptake and migration of substances. A stationary osmotic pressure in the cell is probably without direct significance for diosmotic exchange. Fluctuations of this pressure, however, if they are accompanied by uptake or discharge of water, may have an accelerating or retarding influence because of this water flow (Pfeffer, 1876a:121).

Obviously, for uptake of a substance and all related questions, not only the diosmotic properties of the membrane but the properties of substances found in the solution also are of importance. Chemical transformations of these substances, however they are produced, will be able to initiate, nullify or modify diosmosis.

The ability of the plasma membrane to close lesions immediately makes the uptake of solids into the cell possible; as we know, this has been observed to happen in the plasmodia of myxomycetes and other cells (Hofmeister, 1867:77). Inside the cell, there may be similar processes. At least, according to observations, crystals and grains of starch, for example, seem to be transferred out of the protoplasm into the cell sap and vice versa. But we still need to determine whether such processes play a significant role in the movement of substances inside the cell. In the protoplasm body, which is covered with a cell wall, solids can naturally not penetrate it, and yet one could imagine--though it seems unlikely--that a substance in solution, which penetrates the cell wall, could be changed into insoluble form between the cell wall and the plasma membrane, and would then be abosrbed by the protoplasm in this form.

In the foregoing, I have developed the most important points to be considered when studying the uptake and release of substances in plant cells. Diosmotic movement, which strives to attain a state of equilibriumm, and the disturbance of this equilibrium through metamorphoses,[45] of the diosmosing substances, and finally the specific diosmotic properties of the cell wall and plasma membrane are, in principle, the great driving forces and regulators of the movement and accumulation of substances within the plant. There are no data among all know physiological facts that could not be reconciled with these principles and with the long-known law (Sachs, 1865:388) that depletion and chemical transformations are the causes of the movement of substances is included in the above statement.

When it comes to specific cases, it is true, the data we have relating to the migration and accumulation of substances can generally be causally explained only partially, or not at all. But, on the other hand, no questions have been asked until now that were specific enough to lead to research in this direction. Also, important factors concerning the structure and specific properties of the plant cell have not yet been sufficiently taken into account. I will not speculate to what extent specific cases permit of a definite explanation, and refer you to notes in another paper (Pfeffer, 1876a:111). I believe I need to discuss only a few general basic features here.

All solutes that diosmose through the plasma membrane also pass through a cell wall that is able to imbibe water, but the reverse is not always true. Thus a substance will be able to distribute in the supporting structure of the cell wall, and thus will be able to reach all parts of the plant without ever penetrating into the inside of a cell. Now, if the plasma membrane allows the substance in question to pass through only in specific cells, that substance will only be taken up into those specific cells. The following example illustrates how a substance can accumulate in the cell, or in any other location, when the uniform distribution expected as a result of diffusion and diosmosis is hindered. I have used this illustration elsewhere for the same purpose.

Place a zinc sheet in a cell formed, say, of parchment paper and then immerse the cell in a copper sulfate solution. Eventually, all the copper will be contained in the cell in metallic form. Obviously, this will be so even if at the same time other cells that are filled with copper [*sulfate*] solution, but do not contain zinc, are placed in the same container at the same time. While copper is precipitated, zinc sulfate is formed. The zinc sulfate must finally be equally distributed in the outside solution and the other immersed cells by diffusion and diosmosis. This example shows, at the same time, how a chemical process can cause only one part of the diosmosing combination to be retained.[46]

It is clear that chemical transformation must become total, even if inside the cell chemical transformation produces only soluble products that can again leave the cell diosmotically. As we can infer from certain facts, the kind of processes we have just mentioned, in principle, also take place in the plant organism, but in a much more complex way than in the simple example above. Now if the migration of substances and related phenomena are to be explained causally, we must keep in mind the totality of the processes within the cells and outside them. No less important a factor is the way different cells act together; indeed, perhaps a final product can only be produced in the same cell when cell sap and protoplasm act together. Here we face many different types of difficult, but also highly important, questions that only future research will enable us to answer.

In all processes that take place in the plant cell it must be remembered that the special conditions under which they occur may eventually affect the result in important ways. I will merely show you in a general way how diosmosis in particular can become significant for chemical processes; I will not touch on all aspects of this point. Even as early as 1803, Berthollet (1803), in his classic book, expressed the view that there is total decomposition when there is a reaction that is, initially, only partial and has one of the products removed. In some cases, such a removal will be possible through diosmosis, and the diosmosis will help turn a reaction, which would otherwise be only partial, into a total decomposition. Thus, to give a specific example,[47] oxalic acid in aqueous solution decomposes a small amount of potassium nitrate. If we now assume that only the nitric acid passes through the membrane, then eventually only potassium oxalate will be present in the cell, if there is an adequate amount of oxalic acid and if the nitric acid outside the cell diffuses into a relatively infinitely large quantity of water, or is removed in some way.

It is, however, generally likely that because of the competition of the molecules--to use Pfaundler's (1874) expression - a stronger acid is partially driven out of a salt by a weaker one (though possibly only minimally). One needs no specific illustration to understand how significant such a process could become in the plant cell and be brought about by the plant cell. Let me remind you that certain acids, such as hydrochloric acid and acetic acid, diosmose relatively easily through the plasma membrane. Yet, it is to be expected that other acids, especially those with high molecular weights, do not have this ability. Therefore it is also conceivable that free inorganic acids occur in the plant and are active in it,[48] and such a thing cannot be declared impossible because hydrochloric acid coagulates dead protoplasm, since the effect of the acid might be eliminated by the activities of life, even if the acid enters the protoplasm.

Even the simple fact of being dissolved in water causes some compounds to decompose partially or even totally and it is likely that most salts, though some probably only in infinitesimal amounts, are contained in dissociated form in aqueous solution (Naumann, 1876-77:544). As I already explained above (p. 93), in a suitable precipitation membrane, when hydrochloric acid is removed the relatively strongly dissociating iron chloride is probably eventually transformed into colloidal ferric hydroxide containing hydrochloric acid. Of course, in other cases, the dissociation of a salt may be limited, even if only one of the dissociation products is diosmotically removed--i.e., when the products that remain in the cell accumulate, and the dissociated salt molecules decrease in the process and eventually cease to exist. The acid salts, which I mention because of their reaction with plant cells, are among the substances whose dissociation when dissolved is so extensive that it may even be debatable whether un-decomposed molecules exist in the aqueous solution (Berthelot and St. Martin, 1872:456). Dilute solutions of potassium bioxalate and also of potassium bisulfate quickly redden the blue cell sap of living cells. But, since there is free acid in the salt solution, we cannot infer that there is diosmotic uptake of potassium.[49]

All external influences that accelerate chemical transformations, or in fact cause them, can obviously also become significant for the movement of substances and for diosmotic processes in general. The effect of heat and light on the movement of substances is well known. Since these factors [*Imponderabilien*] increase the extent of movement of the molecules and of their constituent parts, they will first tend to decompose more complex molecules. The degree, and the result, of such a dissociation will be determined in a specific way by the influence of diosmosis. In any case, such a dissociation can become the cause of far-reaching reactions. Thus, as I mentioned before (p. 94), the reaction of chlorine and hydrogen, which takes place in light, is accompanied by an explosion, as a result of the light induced splitting of a number of chlorine molecules. Further, carbon dioxide develops from a mixture of iron chloride and oxalic acid. Also, ferrous chloride is formed, while the split-off atoms of chlorine combine with hydrogen, and the oxygen of the decomposed water molecule oxidizes the oxalic acid.[50] These examples are simply intended to demonstrate that any dissociation, no matter how it is caused and even if it is unproductive in itself, may still bring about far-reaching and complex decompositions. These, under the conditions present in a plant, may prove to be even more complex. Not only are there structurally complex and (as experience shows) easily changeable molecules of organic compounds present in the plant, but in addition there are special circumstances in which the specific diosmotic properties probably also play an important role. With the assistance of diosmosis, not only is extensive decomposition possible through dissociation (which is insignificant in itself), but there is also a possibility that perhaps a dissociation product only becomes effective in another cell. Even if the quantity of the effective substance is very small, this effect may be very extensive. This is shown, e.g., by certain enzymes, which are able to transform a disproportionately large amount of a substance chemically (see p. 94).

It is impossible to predict with certainty whether and to what extent the high hydrostatic pressure in plant cells affects the completeness of a dissociation process. No definitive experiments that could allow us to evaluate the matter have been done. Based on the relation between melting temperature and pressure (Clausius, 1876:172), which has been postulated by theory and confirmed by experiments, we might expect that, in general, the dissociated quantity will decrease as pressure rises, if the dissociation products take up a larger volume than the undissociated substance. While in the opposite case, the quantity of dissociated substance would increase with pressure. Though I have certain reservations, I feel the disintegrations resulting from the dissociation could be linked with other molecular rearrangements, which produce a different result than that which is to be expected normally. By the way, we should probably assume that the pressure, which at a few atmospheres is low compared to the molecular forces, is probably not particularly significant for the dissociation process.[51]

21. Pressure Relations in the Cell.

As already shown, the osmotic phenomena caused by the substances contained in the cell are determined by the plasma membrane, not by the cell wall. This necessarily follows from the consideration that the non-diosmosing substances contained in the cell are in contact only with the plasma membrane, and that the osmosis is produced only by molecular forces effective over an unmeasurable distance. Experience also shows that the turgor of a cell immediately drops if a cell is killed and the continuity of the plasma membrane is disrupted, the reason being that the cell contents produce only very low osmotic pressure across the cell wall (Pfeffer, 1873:140). The potentially very high hydrostatic pressures in plant cells, in spite of very dilute solutions, were very difficult to account for as long as the significance of the plasma membrane was overlooked (as had been happening without exception until now), and as long as the emphasis was on experiments with osmotic pressure carried out on the cell wall, or similarly acting membranes.

As with the porcelain cell in our apparatus, so in plant cells the cell wall forms an abutment for the plasma membrane, which bounds the protoplasm body toward the outside and thus can develop high hydrostatic pressures in conjunction with the resistant cell wall. Where cell sap is present, it is separated by a plasma membrane from the protoplasm. The protoplasm also exerts its osmotic effect at this membrane, an effect which, however, is counteracted by an equal but opposing pressure caused by the osmotic effect of the cell sap. From a purely formal point of view, this box-within-a-box system would resemble a porcelain cell with a surface precipitation membrane inside which a second, smaller cell (with different contents) floats freely. If the content of the latter were to develop higher osmotic pressure, then its precipitation membrane -- presuming that conditions for membrane formation have been fulfilled -- must increase its area until an equal osmotic pressure has been produced, inside and outside of the free-floating cell, by the dilution of the contents and the simultaneous concentrating of the surrounding liquid. Exactly the same is true for the plasma membrane that separates the cell sap and the protoplasm. Also, other structures behave quite analogously -- for example, globules of tannic acid that are covered by a precipitation membrane that is capable of growing - whether such delimited structures are located in the protoplasm or in the cell sap.

What I just said regarding the cell sap and other structures delimited inside the cell by a precipitation membrane so clearly follows from remarks made earlier that it is not necessary to give further arguments in support of it. Incidentally, cell sap would necessarily have to be squeezed through the protoplasm, which is easily permeable to water, by the pressure that an elastic, tight cell wall exerts on the protoplasm body, if this were not prevented by a corresponding counterpressure. Such pressure in the cell sap can obviously only be produced by osmotic action. The substances contained in solution in the protoplasm must obviously produce an osmotic pressure that is defined by their effectiveness inside the plasma membrane. Yet, we must also consider whether and to what extent the protoplasm body can develop pressures other than osmotic pressure, due to its particular structure and properties. Such is actually the case, but these pressures (as I shall show below) are only slight when compared to the hydrostatic pressures and are a factor responsible for the change of form of the protoplasm body, but do not account for the potentially very high pressure exerted by the cell contents against the cell wall. May I preface my remarks by saying that the plasma membrane (which is capable of growth), even if it forms curved surfaces of very small radius, is still incapable of developing considerable resistance to counteract pressure.

Pressure exerted against highly hydrated protoplasm is transmitted quite evenly, as observation shows, and there has probably never been any question that in this respect protoplasm behaves roughly like a viscous substance. For example, the plasmodia of myxomycetes yield to the light pressure caused by a hair, and the fact that they immediately return to their previous shape as soon as the pressure stops indicates that, in spite of the pressure, the effective formative forces in the living protoplasm persist. At the same time, the flow of the protoplasm away from the place at which pressure was applied to places where there is less resistance demonstrates the hydrostatic transmission of pressure. Yet, analogous phenomena may be observed in the protoplasm of other plants as a consequence of any kind of pressure. The tendency to assume a spherical shape, which is generally persistent in protoplasm, shows that not only is pressure hydrostatically transmitted, but also that neither the inside of the protoplasm nor its peripheral covering offer the kind of resistance that would correspond to the state of cohesion of a solid body.

Protoplasm does not possess the aggregate state and the resistivity of a true gelatin, at least not as an entity in itself, as phenomena such as the flow in response to pressure show. But if a relatively more solid substance were to form a fiber network [*Balkennetz*] in the protoplasm, this network, for reasons given above, could not, in any case, have a considerable degree of solidity. It therefore could not, by swelling or in other ways, develop pressures of several atmospheres as actually found in the protoplasm. Here, and in what follows, we are going to consider only the behavior of the living protoplasm body. It would be irrelevant for the final conclusions that will be developed from this behavior if any kind of ground substance of the protoplasm were to have more solidity in itself, but were to yield to every pull and push in the living protoplasm, perhaps for the same reasons as the plasma membrane. It follows necessarily from all the facts put forward here that any pressure in the protoplasm that is at all significant can only be produced osmotically. Solid bodies and bodies capable of swelling will already be able to play a role if they only imbibe water, thus bringing about [*increased*] concentration of the osmotically effective solution. Such solid bodies can also be relevant for local pressures if, for example, they are wedged in between the two plasma membranes at a thinning of the protoplasm layer on the cell wall.

The above considerations are compelling enough, for the conclusions we have drawn, for us to dispense with other arguments. Also, when I allude to the fact that there is a tendency to assume spherical form, I have actually already implied why the state of cohesion of the protoplasm body (and thus also the state of cohesion of the plasma membrane) does not suffice to offer significant resistance, even with very great curvature. All other things being equal, the resistance is inversely proportional to the radius (Pfeffer,

1875d:114). The closest possible approximation to spherical shape, which non-living protoplasm produces, both when it is a continuous mass and when it forms a covering layer around cell sap, proves that neither the protoplasm nor the plasma membrane offer significant resistance to stretching. Otherwise, an equilibrium shape would be reached before the sum of the reciprocal values of the main radii of curvature became the same at every point of the surface, i.e., before the body in this case became a sphere (Wüllner, 1870: vol. I, 275). Incidentally, our final conclusion as regards the resistance to stretching can also be inferred from the expansion and contraction of protoplasm bodies when the concentration of the surrounding medium changes only slightly. Earlier, we showed that for the plasma membrane only a slight stretching force will cause the surface area to grow. The idea that the wall layer of protoplasm, like a vault, might resist a pressure that weighs down on it (or produce higher pressure by swelling) will be immediately refuted by the above arguments when pressure other than osmotic pressure is involved. Incidentally, there are examples where the protoplasm is bounded by flat surfaces parallel to which, or against which, the pressure of the elastically taut cell wall acts.

Slight pressures must, it is true, be created in the protoplasm body other than by osmotic action, as can be inferred from the protoplasm's changes of shape. There may be mechanical causes for some protuberance being formed on the protoplasm body -- for example, a locally increased pressure or, alternatively, a lessened resistance in the peripheral layer (this lessened resistance could be produced by non-uniform thickness, or properties, or by locally enhanced conditions for growth of the plasma membrane or hyaloplasm). What we know about the protoplasm's tendency to assume spherical form makes it very unlikely from the start that, at least in objects immersed in water, unequal resistivity of the peripheral layer would cause protuberances or other changes of shape to appear on the protoplasm body. Locally different resistivity of the peripheral layer might have to be considered for such changes in plasmodia of myxomycetes that are in contact with air. Should this be the case, however, the basic cause of such shape changes in plasmodia would not be the different resistance. This can be clearly inferred from various facts. I shall mention only the outflow of fluid protoplasm out of the more solid peripheral layers, which could be emptied in the process.

I shall not present further facts that also show that

bulges and the like are produced by pressure that has developed in the protoplasm and acts locally. Such a pressure need only be very minimal, for basically it has to overcome only the resistance forces that originate in the peripheral layer of the protoplasm (cohesion, tendency to assume spherical shape). Incidentally, a very small pressure difference suffices to produce protuberances into the cell sap, which is subjected to high pressure -- for an equal osmotic pressure continues to exist on both sides of the plasma membrane. Such pressures may be produced due to the structure of the protoplasm, say, by swelling promoted uni-directionally. There is no contradiction here with our earlier conclusion, which only postulates that higher pressures are caused by osmosis. By the way, even protoplasmic streaming directed against one point will be able to create the energy necessary for some degree of bulging at this point.

Protoplasmic streaming, no matter what its origin, must obviously exert a certain mechanical force (this never amounts to more than a very minimal pressure). The reasoning involved is simple. According to Toricelli's theorem, the height of a column of liquid h, which produces a known outflow velocity v, is given by $h = v^2/2g$, and this height obviously also determines the pressure that would be produced if the entire kinetic energy of a liquid flow was transformed into mechanical energy.

According to a table compiled by Hofmeister,[52] no protoplasmic streaming reaches a velocity of one mm/sec, and if we accept this upper estimate, we obtain h = 1/2.9809 = 0.000051 mm, i.e., protoplasm that flows at the assumed velocity would at the most be able to produce a pressure equaling that of a liquid column of protoplasm 0.00005 mm high, which is extremely low, even if the specific gravity of protoplasm should be much higher than that of water. In fact, the pressure potentially exerted by flow is always even much lower, and therefore there is no point in showing by calculation what pressure, say, a rotating protoplasmic streaming, whose velocity and path curvature are known, exerts against the cell wall. However, it is noteworthy that even pressure produced by relatively slow flow is capable of producing considerable bulges in the protoplasm; such bulges are often observed where protoplasm accumulates because of a very sharp curve in the path of the flow. This clearly shows that an extremely low pressure can alter the form of a protoplasm body.

No matter if protoplasm forms a simple wall layer,

permeates the cell space [*cell sap*] in strands and bands, or has some other form, the principle of the equality of force and counterforce must necessarily always be preserved. Whatever the moving forces and the resistances against them, for a static state to exist, the resultant of the components taken in the three spatial dimensions for each point of the surface must produce a sum of zero. It is not my intention to describe to what extent the mechanics of formative processes in the protoplasm body can be explained according to this principle. I also do not plan to discuss the formative forces themselves and the causes of protoplasmic streaming. At present, this would not lead to satisfactory results.

As far as possible, I intentionally discussed the mechanical resistance and mechanical activity of the protoplasm without a definite model of its structure. As for the aggregate state, it can surely not be that of a liquid, as other authors have already inferred. A liquid transmits a pressure completely hydrostatically. On the other hand, it is necessary to have cohesion that is only minimally different from the cohesion of a liquid in order to be able to exert upon one point the kind of small pressure difference needed to produce changes in the shape of the protoplasm. The aggregate state of a body that is only slightly gelatinous would be adequate in this respect.[53] Such an aggregate state is compatible not only with the sinking of specific heavier bodies (Nägeli and Schwendener, 1867:382), but also with the suspending of bodies that do not deviate too much in their specific weight. Incidentally, this suspension persists for a long time even in viscous slime, and can become constant when processes of movement, as are active in the protoplasm, continue to cause redistribution.

We cannot at present determine conclusively whether the protoplasm body, except for its peripheral covering, is basically a homogeneous mass or is always permeated by a fiber network of ground substance.[54] At any rate, the protoplasm body has a spongy consistency wherever substances that do not directly belong to the protoplasm are distributed in it. But even if we do not decide this question, actual observations show that neither the entire protoplasm body nor a possible ground substance within it can exhibit the resistivity of a really solid body.[55] Of course this does not exclude the possibility that the ground substance may be a solid body in itself, but allows, without difficulty, the displacement of its constituent particles in the living protoplasm, for the same reasons as those that we discussed for the plasma membrane.

Processes of movement may well help us reach conclusions about the structure of the protoplasm, but the structure itself will be linked, in our concept, to a static state in the protoplasm. This factor has often been neglected in discussions on the structure of protoplasm. Another neglected concept is that a change in the form of the protoplasm body is impossible without variation of the resultants of motive force and resistances. Thus a recent study by Velten (1876) did not take into account these fundamental principles when looking at the sperical shaping of the protoplasm, at least as far as I can judge by the note I have read on the subject. As we know for the properties of protoplasm when it has a high water content, the tendency towards a spherical form is always present as surely as molecular forces are active between the constituent particles, but the form of the protoplasm body in any given case is the result of these tendencies and of other forces.

Let us now return to look at pressure conditions inside the cells. Earlier I showed how and why in our precipitation membranes, when solutions had the same concentration, crystalloids produced a completely disproportionately higher pressure than colloids (p. 75). Obviously, in order to produce osmotic balance between substances of the two categories, the solution of the colloid must be made much more concentrated than the solution of the crystalloid. The required difference in concentration depends not only on the nature of the substances used and on the specific diosmotic properties of the membrane, but also on the degree of concentration of the solutions, since concentration and pressure are not in a constant ratio. Starting from a certain density of the solution, the pressure probably, in general, increases faster than the concentration, while below this density exactly the opposite may be true as was shown strikingly by the experiments with gum arabic in copper ferrocyanide membrane.

The cell sap is mostly a dilute solution of salts which, however, has to produce the same osmotic effect as protoplasm, which is rich in substances. Protoplasm contains quite a large amount of proteins. These colloids, judging by their conspicuously slow diffusion, will always produce only low osmotic pressures, as experiments with liquid gelatin glue and conglutin also showed (p. 75). Therefore, it is likely that dissolved proteins, at high concentration, will be contained in protoplasm, even when other substances dissolved in the protoplasm produce an osmotic effect that is not much lower than that of the cell sap. But because we do not know to what extent other osmotically effective substances are found in the protoplasm, it is impossible to infer that the protein substances in protoplasm must form a concentrated solution. By the way, certain observations, such as the phenomena that occur during coagulation, indicate that dissolved proteins are in fact found in the protoplasm in large amounts. Similarly, it now becomes very clear why colloidal tannic acid is contained in tannic acid globules in relatively high concentration (Pfeffer, 1873:12). Vice versa, this fact shows that the osmotic effect of tannic acid across the covering membrane is only small, and also shows that crystalloids cannot be mixed with the tannic acid in any significant quantity. Also, a large admixture of colloids in the cell sap will, perhaps, increase the osmotic effect of the sap only slightly.[56] For example, with copper ferrocyanide membrane, an osmotic pressure of 26cm of mercury was produced by a 6% solution of gum arabic, while a potassium sulfate solution of only 1% produced an osmotic pressure of 192cm of mercury.

With two or more substances that do not mutually decompose, the osmotic effect of a solution will probably not be much lower than the sum of the effects produced by the separate substances in isolation (p. 67). Of course, this no longer holds true as soon as chemical transformations begin to play a role. If, for example, a crystalloid turned into a colloid, the osmotic effect of the mixed solution would probably decrease to a very large extent. I cannot say if this happens, e.g., when certain salts combine with proteins. Whatever the case, qualitative and quantitative data on the ash of the constituent parts of protoplasm do not allow us to infer the osmotic effectiveness of the latter in the protoplasm. Of course, if we had knowledge of all the solutes, their indiidual osmotic effects, and the osmotic effect of the mixed solution, we might be able to draw conclusions about the constitution of the bodies in the solution. Using our osmotic apparatus, research along these guidelines might decide questions of a purely chemical nature.

The contents of a cell naturally exert the highest osmotic pressure against the cell wall when the cell wall only imbibes water. If the water is replaced by a solution, the osmotic pressure exerted upon the cell wall decreases accordingly, and becomes zero when the solution (which the cell wall imbibes) now in contact with the plasma membrane produces the same osmotic effect as the substances contained in the protoplasm. Then, as we know, further concentration of the outside solution contracts the cell contents to a degree where osmotic equilibrium is restored as a result of the increasing concentration of the protoplasm and the cell sap, caused by the decrease in volume. Protoplasm and cell sap, and also those solutions enclosed by a membrane inside it, must always change their volume in such a way that the osmotic forces effective on both sides of the membrane are in a state of equilibrium. But since concentration and pressure are not in a constant ratio, the decrease in volume may be unequal. If there is a highly concentrated solution in the protoplasm, then, according to what has been said, it is quite probable that the volume of the protoplasm will decrease less than that of the cell sap. As the speed of the contraction indicates, water rapidly moves through the plasma membranes, but apparently not more rapidly than through artificial precipitation membranes. At least the microscopically small cells formed from tannic gelatin glue shrink very rapidly in sucrose solution. The completely exposed surface of the membrane will, however, accelerate the water movement (see p. 59), but it must be remembered that in these small experimental objects the ratio of surface area to volume is increasing.

A solution that causes the protoplasm to separate barely noticeably from the cell wall will produce a pressure that is only slightly higher than that exerted by the contents of a cell against the cell wall, which imbibes only water. Here the assumptions are that the space enclosed by the cell wall has remained unchanged when the protoplasm contracted, that both sides of the plasma membrane are osmotically equivalent, and that substances do not diosmose through the plasma membranes in either direction, which is at least very nearly achieved when, for instance, sucrose solution is used. Of course, it is impossible to determine the pressure existing in a cell in this way as long as we do not know the osmotic effect of any substance on the plasma membrane. Therefore it will be necessary to determine this effect for a substance by determining the pressure of the cell contents against the cell wall and the concentration necessary for a sucrose solution, or some other substance, to offset this pressure.

In order to do such a thing, we still do not have enough exact measurements of the osmotic pressure exerted against the cell wall. But once the osmotic effect of a substance on the plasma membrane has been obtained in the manner given above, it will become possible to determine the osmotic pressure existing in the cell by measuring the concentration of the solution in question necessary to initiate the contraction [*incipient plasmolysis*] of the protoplasm. Of course, we then still need to know the relationship between the concentration and the pressure for this substance. Moreover, the plasma membranes of different cells would have to behave identically. If we extended these experiments for determining osmotic effect to include a greater number of cells, both of these factors could be evaluated, up to a certain point.

An equal contraction with the same cell indicates equal osmotic effect for the medium used. Therefore, by determining the concentration of solutions of different substances necessary for an equal degree of contraction, the relative amounts of these substances that produce the same osmotic pressure in the plasma membrane can be determined.[57] But the relationship found in this way will not have to remain the same if the density of the solutions is changed, since concentration and pressure do not increase in a constant ratio, and since they increase unequally for different substances. Now when the series [*of substances*] used in one cell type produces also the same contraction [*incipient plasmolysis*] in another plant, this indicates that it is highly probably that the osmotic effects of these plasma membranes are identical. Incidentally, it is likely that the pressures produced by the same solution with the plasma membranes of various cells will turn out not to be significantly different.

The research necessary for utilizing the insights I have just discussed, in a practical way, has not been done yet. Still, results obtained with artificial precipitation membranes show, when compared to the concentration of a solution needed to abolish turgor in plant cells, that high osmotic pressures can be expected in plant cells. Thus, in the parenchyma cells of Cynara scolymus filaments, initial contraction of the protoplasm body is, on the average, produced by a sucrose solution of 8%-10%, and such a concentration would cause a pressure of about five atmospheres with a copper ferrocyanide membrane.[58] In the petiole joints of Mimosa pudica, the osmotic pressure was normally balanced by a 6% solution. Such a solution, it is true, would not produce a sufficiently high pressure with a copper ferrocyanide membrane as that which must exist in active parenchyma cells of the joint, in which even the partial abolition of the turgor by stimulation can cause pressure to fall by five atmospheres (Pfeffer, 1875d:112).

But the osmotic effect of sucrose on a copper ferrocyanide membrane is, after all, no indication of its osmotic effect on a plasma membrane, which must actually yield much higher values because the high pressure that exists in the joint cells of Mimosa is produced osmotically.

The measurements I made, with plants able to exhibit movement, on the force of these movements could not, by the nature of things, give an exact measure for the pressure existing in individual cells. They only show the minimum pressure that must exist, at least in certain active cells, if the motive force is osmotic in origin. Therefore it is completely inadmissible to compare numbers obtained in this way with the concentration necessary to contract the protoplasm, not even in order to estimate, on this basis, whether the same substance causes equal or unequal pressure with plasma membranes of different cells. Therefore I shall limit myself to the brief remarks above, and merely point out what an enormously high osmotic pressure must sometimes exist in cells. For example, in cells of red beet with a very high sucrose content, initial contraction of the cell content does not occur until the sucrose solution reaches 27%. Elsewhere, I have explained that the cell wall can resist very high hydrostatic pressure, and why.[59]

The form and volume of an isolated cell are always the resultants of elasticity and expandability (or ability to grow) of the cell wall on the one hand, and the pressure of the contents bearing on it on the other hand. If one of these components varies in intensity, the cell volume, too, must change, and all the more so the greater the expandability of the cell wall. But here let us first assume that the volume and form of the space enclosed by the cell wall remains unchanged. This is in order to see the effect that an increase or decrease in the amount of osmotically effective substances must have if only the protoplasm or only the cell sap is affected by it. We shall further assume that no osmotically effective substance leaves the cell or is diosmotically exchanged between the protoplasm and the cell sap, while the plasma membrane itself retains its osmotic properties without modification. Let me again emphasize that a solute produces maximal osmotic pressure in a membrane as soon as the solute does not diosmose through the membrane.

Of course, when there are chemical transformations, the osmotic pressure may change, and it is well-known that chemical processes occur both in the protoplasm and in the cell sap. If we first assume that there is an increase in osmotic effect of the substances contained in the cell sap, then the cell sap will dehydrate the protoplasm until osmotic equilibrium again exists, on both sides of the protoplasm, because of the dilution of the cell sap and the simultaneous concentrating of the protoplasm. While here the volume of cell sap increased and that of the protoplasm decreased, the exact opposite will occur if the osmotic effect of the cell sap is reduced. Pressure increases in the protoplasm must also increase the volume of the protoplasm and decrease the volume of the cell sap, while the lowering of pressure in the protoplasm must lead to the exact opposite result.

In all cases, the pressure exerted against the cell wall rises when the osmotic effect in the protoplasm or in the cell sap increases, and drops if the reverse is true. If it is known whether the pressure against the cell wall is falling or rising,[60] and if the relative volume change of protoplasm and cell sap is also known, it thus can be established in which of these the osmotic effect varied, and whether the variation was positive or negative. If structures (for instance, drops of tannic acid), covered by a precipitation membrane having the ability to grow, are present in the cell sap or protoplasm, any change in the protoplasm and cell sap volume must have a concurrent effect on these structures as well. That is, their volume falls as the surrounding medium becomes concentrated and rises when it is diluted. When such a volume change takes place, and when it is also known either how the pressure against the cell wall or how the volume of cell sap and protoplasm is changing, then in both cases we have discovered where and in what sense osmotic variation occurs, provided that the osmotically effective substance remained the same in the aforementioned structures themselves.

It is easy to deduce from the above how circumstances must develop if a fluctuation in the osmotic effect appears simultaneously in the protoplasm and the cell sap. According to the rules of probability, a volume change can almost always be expected, of course, as a resultant of the two forces, which act in the same or in opposite directions. If there is an actual volume change, it is obviously impossible to know if the change comes from the protoplasm only, from the cell sap only, or from both at the same time. This cannot be decided even if, at the same time, we know the change in pressure exerted upon the cell wall and the change of volume ratios in other structures covered by a precipitation membrane (tannic acid globules, vacuoles, etc). At least a definitive final conclusion could only be drawn if there were specific quantitative data available, and at present it seems to be impossible to obtain them. Of course, if for instance the pressure against the cell wall increases and simultaneously the protoplasm volume increases, this is a clear indication that the osmotic effectiveness of the protoplasm has risen, at any rate. Yet it will still be doubtful whether during this time there was a lesser increase or decrease of osmotic effectiveness of the cell sap. There will still be the same question in all other cases, e.g., when the cell sap gains or loses volume while the pressure against the cell wall rises or falls. But if by chance the osmotic effectiveness of the protoplasm and cell sap were to rise in the same ratio, so that the two volumes remained constant, this increase in pressure would be a result, on the one hand, of pressure against the cell wall and, on the other, of the volume decrease of tannic acid globules or other enclosed structures. Simultaneously, it becomes certain that there has been a proportionately equal change in the osmotic effect of the protoplasm and cell sap.

The result of osmotic pressure fluctuations in cells that are covered with elastic and expandable cell walls, and in cells which have no cell wall, is easy to predict. If the osmotically effective substances remain completely unchanged, while the volume of the cell decreases in response to increased pressure from outside, then protoplasm and cell sap will both take part in the release of water in essentially the same way, just as they would during contraction, where a dehydrating medium is involved.

If, contrary to our current assumption, the diosmotic properties of plasma membranes were to change in such a way that the same solution now produces a different pressure in the plasma membrane, then there could also be fluctuations of the relative volumes of protoplasm and cell sap, whose course is easy to predict, under the given conditions. The plasma membrane could cause the osmotic pressure to rise or to drop in two ways: either by allowing a substance, which did not diosmose previously, to pass through, or by modifying plasma membrane properties, without such a change [*substance remains impermeable*], in such a way that the same solution now produces a different pressure, no matter how such a modification was brought about. To make the pressure rise or fall significantly in the latter way the composition of the membrane, at least, would have to be modified in a significant way (some kind of infiltration would hardly suffice). The properties of the plasma membrane known at present -- its diosmotic behavior, the way it is produced and continues to form -- do not point to such a modification at all. But when the diosmosis of a substance that did not previously diosmose is initiated -- if such a thing can happen at all because of changes in the membrane -- then the pressure produced by the non-diosmosing substances at first remains unchanged. Furthermore, when there is only slight exosmosis of a substance, the pressure deviates only slightly from the maximal pressure level that the same substance would produce in the same membrane if diosmosis stopped. The slow exosmosis of the [*osmotically*] effective substances can be deduced from the fact that such substances contained in living cells, in the form and associations in which they occur in the cells, do not diosmose through the plasma membrane in significant amounts, as experience shows. Thus it appears likely that changes in the plasma membrane do not produce very significant pressure fluctuations. Of course, I cannot at present offer a compelling argument to support my statement. Let me also emphasize that the thickness of the membrane does not affect the pressure.

There can be no doubt that the logical conclusions developed on the basis of the properties and osmotic effects of the plasma membrane will be expressed, during the life of the plant cell, in the relative volumes of the protoplasm, the cell sap, and the structures enclosed by the plasma membrane. This will, however, often be complicated by changes in the total volume of the cell and by other circumstances. For example, if starch or oil is formed from glucose, then, if this process takes place in the cell sap, the osmotic effect of the sap must necessarily decrease, possibly even very greatly. Processes of similar significance osmotically, for cell sap or protoplasm, could also be mentioned, based on known examples. Also, we actually know, of course, that the water content of protoplasm is apparently different during different phases of development of the living cell. According to what we have stated, this must necessarily be the case under certain conditions. According to the above, and according to all we know about the way the plasma membrane is formed, we can also predict under what circumstances vacuoles must appear in the protoplasm, and when and how the vacuoles become larger. True, we will still need more extensive research to learn to reduce these investigations and other questions to their causal basis in a more definitive way. On the other hand the insights we have gained can be used in combination with other facts to follow closely the kinds of chemical transformations inside the cell.

We will now also be able to attack, with new tools, the

processes of movement and growth that occur more rapidly (irritability, periodic movements, etc.). We will be able to discover if the cause for movement lies in osmotic pressures and, if the answer is affirmative, we can determine if the relevant effect takes place in the protoplasm or in the cell sap. From the relative volumes of the latter, we could determine this only with the reservations indicated earlier. Of course, in practice, there will be considerable and possibly insurmountable difficulties in determining such relative volume changes, even when the volume and the form of the cell are kept approximately constant and the experimental material does not present other difficulties with respect to measurement. After all, even if the protoplasm forms a simple wall layer, because of the changes in shape of this body it will not be easy to check the volume of protoplasm and cell sap with sufficient accuracy by direct axial measurements. It is even less likely to expect favorable results when the protoplasm body has a more complex shape.

At any rate, such axial measurements can only be considered when volume fluctuations are very large. This can be expected in certain stimulation and periodic movements, where pressure may drop by more than half, i.e., by many atmospheres. In order for their osmotic effect to be reduced to that extent, inherently dilute solutions such as cell sap must at least be further diluted by water uptake, perhaps proportionally to the reduction in pressure. In the case of concentrated solutions (of the type that perhaps exist for colloids in the protoplasm) we must consider that, starting from a critical concentration level, pressures can increase faster than concentration.

But, as I showed earlier, in order to examine the questions that preoccupy us here, we can also use structures covered by a membrane capable of growing, of the type present in vacuoles and tannic acid globules, and perhaps also in the cell nucleus, the chlorophyll granule, and so forth.

In such structures, if they keep their spherical shape, a volume change can be checked more accurately and with greater hope of success, since only two axial measurements need be made. We can expect success if the volume fluctuates by 1/4 or more, although of course the radii increase or decrease only in the ratio of the cube roots of the volumes. The volume of such structures changes according to the osmotic effect of the medium that surrounds them and, when such fluctuation occurs, it is also possible to obtain definitive answers to our questions if the volume of the cell increases or decreases at the same time. But if, for the structures in question, volume changes had actually been calculated when the cell volume was constant or variable, it would be possible to calculate the amount by which the pressure itself had changed. Further conclusions could also be drawn in conjunction with other observations, and placed on a sound basis.

Probably what usually causes the fluctuation of osmotic pressure is not modifications in the plasma membrane but processes in the cell contents. We can refer to the totality of these as chemical processes, since they always involve changes in the properties of a substance. Generally, any chemical reaction, if it involves osmotically active substances, will cause the pressure to drop or to rise. For, a substance is not likely to have the same osmotic effect as the products derived from it. With metabolism, the osmotic pressure in a cell must very often fluctuate and disruptions that affect metabolism also affect osmotic pressure. It is obvious that osmotic pressure must drop when, e.g., oil or starch is created from glucose. Also, when a colloid is produced from a crystalloid, the sometimes enormous difference in the effectiveness of the two substances will probably cause very great differences in osmotic pressure without an insoluble substance being formed.

Thus the causal explanation of pressure fluctuations will normally be linked to the clarification of chemical processes for whose beginning and development the specific structure of the plant cell is very important. For instance, I already used one example, iron chloride, to show that with the help of diosmosis a colloid may be formed from a crystalloid (p. 93). Osmotic processes are no less involved in such occurrences, though these occurrences affect osmotic pressure only indirectly, i.e., have a releasing effect on osmotic pressure or on the substances that cause osmotic pressure.[61] As ferments show us, a small amount of a substance can potentially help bring about the chemical decomposition of a large or even unlimited quantity of another substance. We have analogies that help us explain the chemical processes taking place in the plant. In almost all cases though, we still do not have an exact explanation for the causes of the chemical transformations actually occurring there, and the transformations themselves are barely understood. A sudden lowering of osmotic pressure, which occurs, for example, in stimulation movements, requires that a reaction occur rapidly. The same type of reaction may be found outside the organism, occasioned perhaps by an initiating process (some small stimulus from the outside) that is infinitesimal in terms of energy required. Examples include the almost instantaneous precipitation of solutes from super-saturated solutions and the sudden transition of dissolved colloids to a so-called pectin-like state -- processes that must necessarily cause osmotic pressure to go down.

Where osmotic pressure involves repeated fluctuations (as in stimulation movements), this pressure must be regenerated either by the intervention of new substances or by the reappearance of the same substance at its point of origin. The latter probably takes place, along with other events, in the case of irritability. But in this case as well, two antagonistic and very distinct activities must be at work. This happens also even if, along with other factors, the molecular forces inherent in matter should bring about a return to the original state, which itself could be disturbed only by another intervention. Even inside the organism, no regeneration of substances can possibly come about other than through forces that act in the opposite sense. If we keep in mind, as we should, that this is not a simple, but a double reaction, then in the majority of cases there are ways of again reversing a chemical process outside the organism.

Regardless of whether the reaction that regulates the osmotic pressure takes place in the protoplasm or in the cell sap, the questions with which researchers must approach these processes remain basically the same. Now, chemical processes, as we know, take place both in the cell sap and in the protoplasm. Therefore we cannot consider a priori that it is likely that the underlying cause of osmotic pressure fluctuation is to be sought, say, in the protoplasm. For obvious reasons, we cannot make such an assumption even if the view that regards the protoplasm as the actual living part of the organism is well-founded.

Fluctuations and effects of pressure, the visible and measurable symptoms of osmotic action, can, it is true, indicate where a chemical process is occurring, but will never directly give more detailed information on the substance, which is changing chemically, and on what type of reaction is taking place. However, in conjunction with other facts, volume and pressure fluctuations will provide points of reference so that chemical transformations inside the cell, and their development, can be more closely followed. It is true that physiological research can already explore areas that have seemed, for the most part, inaccessible until now, by focusing only on pressure and volume relations. For example, research on the cause of movement processes never got beyond the question whether the determining factor was a change in the cell wall or in cell contents. At present we have reason to hope that we can successfully study the morphologically structured cell contents in a physiological sense as a complex osmotic apparatus by looking at all the questions that are related to osmotic exchange and osmotic effects.

Volume and pressure changes are always only the consequence of chemical processes (or physical processes). Any inquiry into the latter is closely tied to the former. If this inquiry cannot immediately be fully answered, then at least any observable reaction can clarify either the chemical process itself or, at least, can be valuable as a way of monitoring osmotic and other processes. Color change, precipitation of substances, and phenomena related to solubility are, among other things, microscopically observable reactions. They could serve to explain chemical processes, whether conditions for such reactions and other reactions are already given in the cell, or whether they still have to be created by preliminary maneuvers. But then, microscopically observable reactions are by no means the only ways of observing processes that develop in the cell.

Whatever methods are appropriate now and in the future, their goal is to attempt to find the causes underlying metabolism and the activities of the cell. In the organism, we have an association of cells whose activities differ in degree and specificity. The metabolism and activities of the individual organs and of the entire organism are the resultants of the processes and effects of the individual cells. Obviously we need to take into account inhibition and resistance by passive tissue, and other circumstances. All chemical and physical processes in the plant are included in the broader concepts of metabolism and energy transduction (activities in the organism). Since we are basically dealing with the effects of molecular forces and the movements produced by the latter, we may speak of the mechanics of the cell, whose goal is to explain the cause of metabolism and energy transduction.

22. Cell Mechanics of Movements.

In growth and in movement, in sap outflow, and generally in the most diverse processes, osmotic pressures play a more or less prominent part. That is why the preceding elementary research is important for a large number of physiological questions. It is true that more extensive work will be needed to utilize fully the general concepts that have been arrived at. Though there are no conclusive studies in this direction available to me yet, I still feel I should examine several physiological processes related to osmotic pressure in order to clarify the present state of affairs. I shall begin with stimulation movements, which provided the first impetus for my osmotic studies when I realized I could not delve deeper into the mechanics of the cell without them.

My first studies (Pfeffer, 1873) of the filaments of Cynareae and of Mimosa pudica joints showed that when there is a stimulation movement, a large amount of water leaves the cells (of this material). This release of water could be immediately perceived on the cut surface of Mimosa joints and Cynara filaments that had been previously injected with water. For the filaments of Cynara, however, the fact that there was release of water was also established by the resulting decrease in volume shown by measurements on the cylindrical cells. During the stimulation movement, as I also showed, the pressure exerted by the stretched elastic cell wall on the cell contents does not increase. Therefore release of water can in no case be caused by a rise in pressure exerted by the cell wall upon the protoplasm body.

On the basis of these facts, I was able to show that it was highly probable that not the cell wall, but the cell contents undergo some kind of change that causes liquid to be pressed out of the cell by the pressure of the elastic and stretched cell wall (this has now been fully proved by our research on osmosis). For even if the cell wall should change in such a way that the same solution would now produce a lower osmotic pressure -- and we were able to make this objection previously — that fact would never cause water to be released from the cell as long as the cell wall exerts pressure on the contents with unchanged force, because osmotic pressue is determined by the plasma membrane. Also, this pressure must remain constant under such circumstances just as much as it would in the same precipitation membrane that is deposited on porcelain cells of varying porosity.

In stimulation movements the function of the cell wall is only to press water out of the cell contents, like an elastically taut balloon, when the pressure inside -- and we are speaking of osmotic pressure — falls. Water is released until, as the cell size decreases, the decreasing elastic tension of the cell wall and the osmotic effect, which increases with the concentration of the cell contents, are balanced. Subsequently the cell is then again stretched back to its former size by the gradually rising osmotic pressure. In the sensitive cells of Mimosa pudica, these fluctuations in pressure may exceed five atmospheres (Pfeffer, 1875d:112) and in the filaments of Cynara scolymus also, these may be above, perhaps considerably above, one atmosphere.[62] But in all cases, osmotic pressure falls only partially during stimulation movement, as the further decrease in size of the still expanded elastic cell walls shows when turgor is nullified. That is why the plasma membrane always adheres closely to the cell wall, and it will be impossible for it to separate from the cell wall even when the stimulated cell is prevented from decreasing in volume. Incidentally, this was also confirmed by direct experiments with filaments of Centaurea jacea (Pfeffer, 1875e:290).

For various reasons, there was no consideration of an increase of osmotically effective substances in the fluid that saturates the cell wall, and thereby causes a reduction of the stimulation. I shall simply point out that Cynareae filaments injected with water and immersed in water may be repeatedly stimulated with the same results. Under these circumstances, however, a good portion of the soluble substances present in the cell wall would surely have been removed. From these and other experiments it also follows that significant quantities of the substances contained in the cells do not, in any event, diosmose through the plasma membrane during stimulation movements. I will just mention something that is easy to observe -- that variable capillary tension of the liquid at the cell wall surface adjacent to spaces containing air can have no significance for water release as a result of stimulation.

The immediate goal of my earlier research had to be to decide whether the cell contents or the cell wall was the part that responds to stimulation and, as I was able to predict, this question has now been definitively decided, since we know the osmotic effect of the plasma membrane. It must be empirically determined how pressure fluctuation comes about in the cell contents; this problem no longer seems insoluble. Lacking the basic data we have now gained, my earlier speculations did not stand on a clear and reliable foundation. My general conclusion though was correct, that stimulation causes changes in the cell contents that produce the release of water brought about by pressure exerted by the cell wall (Pfeffer, 1873:155) I still was on the wrong path when I tried to demonstrate that stimulation movement takes place when molecular interstices in the plasma membrane (which I called a primordial utricle) expand.[63] This view is refuted by the fact that during a stimulation movement the substances contained in the cell are not diosmotically removed from the cell, at least not noticeably, and under these circumstances pressure cannot be lowered by the expansion of the molecular interstices in the plasma membrane. This also cannot occur by means of exchange of osmotically effective substances between the protoplasm and the cell sap.[64] As a result of such a process, pressure could, of course, drop only if it was reduced faster than the concentration when the protoplasm or the cell sap was diluted (see p. 64). Now, the large pressure fluctuations must be based on osmotic forces since, as I showed earlier, protoplasm is incapable of developing any significant pressure (for instance) by acting like a swelling solid.

There are two other possible ways stimulation movement could be caused. Either osmotic action in the cell sap or in the protoplasm must drop or change simultaneously in both, or else the plasma membrane must undergo such modification that the same substances, without diosmosing, now produce a lesser osmotic effect. The latter situation could come about, for instance if the diffusion zone was extended (p. 50). While it is not exactly possible to refute the latter alternative at the moment, that alternative seems quite unlikely, since the molecular effect created by the plasma membrane would have to be altered significantly to reduce, for instance, the pressure in Mimosa to less than half (decreased more than five atmospheres). This cannot be assumed if we consider that the other diosmotic properties of the plasma membrane remain the same during stimulation movement, as far as we can judge.

Therefore it seems fairly certain that the actual cause of pressure fluctuation is a change in the cell contents produced by stimulation, which reduces the osmotic effectiveness of the cell contents.[65] Whether this process takes place in the protoplasm or in the cell sap, or in both, can only be decided empirically. I hope it will at the same time be possible to reach a definitive conclusion with respect to the altered osmotic effect of the plasma membrane.[66]

Since we are focusing only on the fundamental mechanical processes in the cell in response to stimuli, I refer the reader to my previous studies and will not refer here to such topics as the detailed description of how stimulation movement takes place in our experimental objects, what path is taken by the water that is released, and other questions. Even though at present osmotic processes have not yet been localized inside the cell, and the immediate causes of osmotic variation can be described only generally as chemical processes (molecular rearrangements), we can still get an idea about how these molecular rearrangements are initiated in the osmotically effective substances, on the basis of considerations I already developed in my work on irritability (Pfeffer, 1875d:142).[67]

If the condition where there is response to stimulus is interrupted by chloroform or constant mechanical shaking, after there has been stimulation movement in a cell, the cell again takes its former shape by returning the osmotic pressure to the same level where it was before the stimulation occurred. If one discontinues the chloroform or stops the shaking, the organ is again able to react. The reaction is not immediate but gradual, so that stimulus first produces only a limited amount of movement, and the full range of movement is only produced after a while. As these observations show, stimulation in the cell only temporarily produces a molecular rearrangement which, independent of its sensitivity to stimuli, again returns to the state of equilibrium that existed before the stimulation. No other interpretation is possible, since the osmotic action in the sensitive and in the non-sensitive cell is the same, and sensitivity to stimuli returns without changes in the osmotic pressure.

The above facts indicate that to restore the sensitivity something has to be added which does not significantly affect osmotic action. The simplest assumption is that a small quantity of a substance is formed that has a triggering effect because it decomposes when stimulated. That is, it initiates the transformation that causes the lowering of osmotic pressure. The return to the earlier osmotic pressure could then perhaps be only a consequence of molecular forces inherent in matter. However, the initiating process does not work like a spark that causes the whole mass to explode when it comes in contact with gun powder. Rather, the triggering force can always transform only a limited, though perhaps relatively very large quantity, of the osmotically effective substance since, as I already mentioned, when sensitivity to stimuli returns gradually then stimulation produces only a sub-maximal amplitude in the movement.

The view that we have derived from our deduction is a hypothesis, but it is consistent with the data we have. Rapid decompositions after impact, even after a slight impact, are sufficiently familiar to us. We could compile data for both the initiating process and the subsequent process that could illustrate purely formally the lowering of osmotic pressure following an impact (see p. 191). I refrain from filling in these details here, and will not explain how the questions that suggest themselves here may be decided experimentally.[68]

No matter which movement processes we examine or try to compare with each other, some of the same questions will always present themselves. If we disregard external or internal impact, and the form and duration of the movement, the first thing we will need to decide will be whether the cause of movement is in the cell wall or in the cell contents. Assuming, for example, that it is osmotic pressure conditions that cause the movement, two different movements might possibly share this immediate mechanical cause, but differ from each other. They could differ regarding the place where pressure originates, or which of the substances is osmotically effective, or at any link in the chain of processes. This chain extends from effects produced directly by an external or internal impact through the processes that eventually regulate the osmotic pressure. Strictly speaking, the cell mechanics of two movement processes can only be called qualitatively identical when all the links in this chain are identical. Yet, even when this is not so, individual processes can obviously still be qualitatively identical or only analogous. The cell mechanics of movement processes must be evaluated according to such principles, both when comparing movement processes in the same and in different experimental objects. The cell mechanics of two movements, where the cause of stimulation and the regularity of appearance are identical, may be completely or partially different, while in another case perhaps the opposite will be true.

There are too many gaps in what we know of the internal causes of the processes of movement and growth, so that we cannot make a comparison of the cell mechanics of different objects. If we focus on the stimulation movement produced by impact, then judging by the discharge of water from Berberis filaments,[69] which takes place when there is movement, and by the relaxation in Oxalis joints (Pfeffer, 1873:74), we can expect that also in these objects the immediate cause for the movement is fluctuations in osmotic pressure. This is probably true of other analogous stimulation movements as well; at least we know of no fact that would contradict it. I would not like to assert, however, whether all of the cell mechanics connected with movement always runs its course in exactly the same way. In the case of tendrils which respond to stimuli, we do not know if it is fluctuation in osmotic pressure that causes movement. However, there are reasons to think this is likely. Growth as a result of movement would then enter as a new factor. Here, however, excitation is not produced by a single impact, but by continuous contact, and we need specific research into this release process. Perhaps this and other effects of contact, as they relate to growth phenomena, can be reduced to uniform points of view.

From Darwin (1876) we have also learned about two distinct movement processes in a number of so-called carnivorous plants. These processes may be called stimulation and resorption movements (Munk, 1876:97). The former is caused by mechanical impetus, while the latter is produced by chemical action, and both may run their course significantly differently. For example, in Dionaea a stimulation movement causes a leaf to close within a very short time, while a resorption movement does not close it until one or two days later. The manner and means of excitation and development of both movements are also different in other ways. However, with the studies available to us now we cannot evaluate to what extent the mechanism of both movements is the same or different. We also do not know whether the cell mechanics of stimulation movements in those plants is the same as that in Mimosa and Cynareae, which seems probable in the case of Dionaea at least (Munk, 1876:121).

Darwin himself did no studies of his own on movement mechanics, and his discussion of the subject is very inadequate. Darwin (1876:234, 289) believes that the cell walls probably contract, that in the process some of the enclosed fluid is squeezed outward and, if this is not so, he states it is most likely "that the liquid content of cells contracts because of a change in its molecular state, followed by the coming together of the walls." Darwin goes on to say that "whatever the case, the movement can hardly be ascribed to the elasticity of the walls combined with a pre-existing state of tension." By this statement he completely contradicts his own view on the stimulation mechanism, at least if we use the terms "elasticity" and "tension" as they are commonly used in physics and physiology. Incidentally, Darwin's statements leave somewhat vague the mechanical processes during stimulation movement. The Darwinian view we first mentioned would agree with the stimulation mechanism as we have demonstrated it in certain types of experimental objects if the lowering of osmotic pressure were indeed the cause for the contraction of the cell walls. I need not criticize Darwin's arguments any further, since Munk (1876:111) has already explained their flaws and inadequacies.

In some glandular hairs, Darwin noticed that under certain circumstances a substance was secreted. True, this is not the immediate mechanical cause of the resorption movement in Drosera and other objects, but may possibly have some connection -- though only a complex connection -- with the movement processes as an initiating process (see Darwin, 1876:208 and Munk, 1876). I do not intend to explore whether this is so, and how it happens, since I intend to deal with the phenomenon only because it is a graphic and informative illustration of certain problems in cell mechanics. I shall speak only of the glandular hairs of Drosera rotundifolia, which Darwin, too, had mainly observed.

In the parenchyma cells in the stem of glandular hairs of Drosera, protoplasm and cell sap are, as usual, separate. Because of dissolved pigment, the cell sap is usually colored red, but it is also colorless in some hairs. The colorless, vigorously flowing protoplasm simply forms a wall layer, or permeates the cell space in strands and bands. Its cell nucleus, in contrast with Francis Darwin's (1876:310) express assertion, is quite normal; it is no more difficult to find than in a number of other parenchyma cells, and cannot be missed once stained with iodine. The precipitation of a substance in the cell sap, as previously mentioned, which is transmitted from cell to cell in the stem of the hairs, can be produced by mechanically stimulating the glands, and also by the action of reagents. I have nothing of importance to add to the detailed description by Darwin (father and son) as regards the precipitation process itself. I refer the reader to the works of these scientists, noting only that the cell sap first becomes cloudy, then the precipitated substances gradually aggregate into larger masses, until finally a number of globular bodies have formed.[70]

As the colorless globules that are formed in colorless cells indicate, pigment is not necessary for the precipitation process. In the pigmented cells on the other hand, pigment is accumulated in the precipitated bodies in a similar way as it would be by other insoluble proteins -- incidentally, the precipitated bodies consist of proteins.[71] Generally, the cell sap loses color completely, but when the cells are very rich in pigment, the sap may occasionally retain a faint tinge of color, apparently because the amount of precipitated material was not adequate to bind all the pigment. During and after this precipitation, the protoplasm body is essentially unchanged, and the uncolored protoplasm is still found flowing vigorously, as before. The flow stops when a certain amount of ammonium carbonate penetrates the cell, but returns some time later if the salt is then rinsed out. If sucrose solution is used for contraction, it is easy to satisfy oneself that the protoplasm body remained always like a sac pressed against the cell wall, as demanded by the circumstance (at least with regard to the plasma membrane) that there continues to be turgor in the hair cells even after the precipitation processes occurred in the cell sap.[72]

The above remarks, and the accounts of both Darwins show that there is no question that this clumping takes place in the cell sap. When, however, the Darwins refer to these bodies as protoplasm, they falter as regards the concept that was intended for this word, and maintained thereafter, by its author. We use the term protoplasm to designate a morphologically distinct part of the cell body, but do not use the word as a collective term for proteins. When Schleiden wanted to adopt the term "protoplasm" in this latter sense, Mohl (1855:690 footnote) himself strongly protested. F. Darwin (1876:315) has again made this error, criticized by Mohl, when he claims that since the globules are composed of proteins, they must consist of protoplasm. It is enough to point out this fundamental error. I can therefore refrain from refuting other equally invalid reasons for which Darwin characterized the precipitated bodies as protoplasm.

The precipitation we have been discussing is caused by mechanical and chemical stimulation of the glands, as we know from C. Darwin. From the glands, it is then transmitted from cell to cell in the hair. As far as we know, mechanical stimuli can act only through the intervention of the gland, but ammonium carbonate causes clumping even without such intervention, which Darwin does not mention. If a single hair with its gland removed is placed in an ammonium carbonate solution (I used a solution of one part of salt to 150 parts of water), precipitation begins very soon in the cells adjoining both cut surfaces, and continues from these along the hair. Within half an hour, precipitation is usually almost complete in all the cells, while in pure water hairs with their tips removed exhibited no change in the cell sap either after half an hour or even after a longer time.[73] The precipitation begins and continues as indicated because the ammonium carbonate easily penetrates the cells through the cut surfaces, while the cuticularized surface of the hair does not allow this salt to diosmose, or does so only very slowly. If the cuticle is injured in any place, the precipitation that originates at that spot immediately reveals ammonium carbonate has entered. Incidentally, a very analogous process, where there is uptake of ammonium carbonate, may be observed when cut filament hairs of Tradescantia are placed in the above-mentioned solution. The color change caused by the penetration of this salt is then transmitted from the injured cells down the rows of cells. The fact is that in the uninjured hairs of Drosera it is the gland at the tip of the hair that takes up the ammonium salt. This is why precipitation would proceed from here to the base of the hair even if the gland was involved only in uptake and in no other way. There are small glands along the side of the hair as well. They take up ammonium carbonate, though apparently more slowly than the main gland, for the contents of these and probably also of the immediately adjoining cell can already contain the globular bodies before the precipitation, which proceeds from the tip of the hair, has reached the point in question.

Thus, after the gland has been removed, it is not a mechanical stimulus but ammonium carbonate that can bring about precipitation in the cell sap. The question now is whether the ammonium salt, which indeed easily penetrates the cell, directly causes precipitation, or whether, perhaps, decompositions are caused in the protoplasm (or cell sap), which have a triggering effect in the cell sap after the products of the decomposition have entered the cell sap diosmotically.[74] At this time, I am unable to decide which of these alternatives applies, but I am convinced that it will be possible to answer the question definitively by means of experiments. Note, by the way, that because of the high molecular weight of the proteins, a large quantity of them can potentially be precipitated from a solution by a small amount of ammonia. In order to determine how and by what means the glands act, for example, when there is a mechanical stimulus, it would first be necessary to study the effect of certain substances on intact hairs and on those that have the gland removed. I have not done so yet. It is, however, noteworthy that dilute hydrochloric acid initiates the dissolution of the bodies precipitated by ammonia, starting at the cut surface of the hair.

Let us look at the way that precipitation proceeds from cell to cell beginning with the mechanically stimulated gland and, quite analogously, from any place where ammonium carbonate had entered. There is no question that as a result of the stimulation some solute diosmoses from the gland into the adjoining cells. This must be caused by an increase in the permeability of the plasma membrane of the gland cells or by a chemical transformation in the content of the cells. The latter seems more probable to me. But then, we would have a case where decomposition is caused by a mechanical impetus. Such a decomposition also yields a substance which, by diosmosing into other cells, affects them, i.e., it brings about the precipitation of proteins in the cell sap. Here I would hope that all the steps in a reaction that can be brought about by a mechanical impetus can eventually be explained. The reaction could also bring about fluctuation in osmotic pressure, but obviously does not have to do so. For instance, while colloids are precipitated, crystalloids that, for example, were previously held in a colloidal compound might simultaneously be liberated and might now, thanks to their greater effect, even make the pressure produced by the resulting solution, with less solute mass, rise. In fact there would need to be some kind of compensation in Drosera if the osmotic pressure were to remain constant, in spite of the massive precipitation of proteins, though of course these are colloids with little osmotic effectiveness. The precipitation we are speaking of can actually take place in Drosera without the hairs being curved inward, but based on this fact we cannot state with certainty that the osmotic pressure is not altered by the precipitation at all.

This initiating process raises a whole series of questions

that seem to be open to experimental treatment. For example, we would have to decide whether the substance released by the glands acts only directly, in such a way as to produce precipitation in the cell sap, or whether the parenchyma cells of the hairs also produce amounts of the substance that can cause the reaction. This is an important question for theoretical reasons. In the first case, the amount of substance that will be precipitated depends on the quantity of reagent secreted by the gland, but not in the second case, where there is even a possibility that the reaction, avalanche-like, increases from cell to cell, i.e., as the amount of available triggering substance increases from cell to cell. For example, the intensity of excitation grows during transmission in the nerve. In Mimosa pudica, transmission from leaflet to leaflet is generally accelerated if the terminal leaflet of a pinnae is stimulated. Here the water flow in the vascular bundle has a triggering effect.[75] Its kinetic energy increases from joint to joint until finally in the primary petiole and in the twig, that is, on entering a wider and longer flow channel, the speed and (because resistance predominates) the kinetic energy of the water flow decreases. It is not possible to determine what happens in Drosera based only on the beautiful studies of C. Darwin (1876:208) on the transmission of the precipitation in the cell sap. One will, incidentally, be inclined to assume that at least in a certain sense, the release process grows in intensity; because the stimulation of a gland may transmit precipitation along the hair and the leaf parenchyma to the upper end of another hair, and thus give rise to the precipitation of a relatively large quantity of proteins.

It may also be possible to discover the nature of the substances that decompose in the gland, and to clarify the process of decomposition itself in its entirety, or at least up to a certain point. Thus, for example, a comparative study of the effect of certain substances on hairs whose tips have been removed, and of intact hairs, and observation of the secretion process, which is said to produce a fatty acid (C. Darwin, 1876:79), can give us important insights. Of course experiments and interpretation need great care, and we must keep in mind that the plasma membrane may separate two substances thanks to its specific diosmotic property. This is also a factor when interpreting the effect of different ammonium salts on hairs whose tips have been removed, since the ammonium salts may be contained in solutions in a partially dissociated state (Naumann, 1876-77:546).

The general insights we gained from studying the

stimulation movements are also true for other morphological changes of the cell, inasmuch as growth processes and movements of organs result from the activities of individual cells. Where the cell's activities are expressed outwardly, and when the cell undergoes morphological changes, pressure from inside and the resistance of the cell wall are always important factors. When the morphological change cannot be reversed we speak of the growth of the cell wall. This growth, when it involves the insertion of new particulate matter, constitutes a separate chapter of cell mechanics.

Sachs (1874:744) made it clear how, and to what extent, growth and dynamic activities, and thus cell mechanics as well, are accessible to experimental research. Like the entire living organism, the cell has hereditary properties that those who study cell mechanics are unable to study at this time. Nor is it necessary for them to do so if they see their goal as reducing the actual activities to their immediate causes (as they occur under the given conditions) and to explain them causally in this sense. This task is analogous to trying to explain the activities of a complex apparatus by looking at the fitting together of all its parts without asking to know how this apparatus came to be constructed.

The work performed by the organism comes about when potential energy [*Spannkraft*] is converted into kinetic energy, and no matter how this transduction takes place, specific molecular processes are always necessary. It is the goal of the study of cell mechanics to investigate these processes. The molecular processes as a whole fall into one of two concepts, "metabolism" or "transduction". By the latter we mean the dynamic activities, while by metabolism we mean all molecular transformations or, to put it briefly, the statics of molecular processes. An "initiating" impetus is any impulse that induces the transduction of potential energy into actual energy.

Just as we distinguish external and internal causes of growth (Sachs, 1874:744), we must distinguish external or induced, and internal or autonomous, causes of initiation. These depend on whether the initiation is caused by an external impulse, or the cause of the initiation has its basis in the hereditary development of the organism and, accordingly, occurs in a particular phase of development. When an initiating process is induced, the initiating force is defined, and in a favorable case we can observe the totality of molecular processes that are linked to the initiation and its development. On the other hand, an autonomous initiation based on a hereditary course of development and formation could only be traced back to its first step by means of a causal explanation of this specific course of development. This does not exclude that a specific chain of mutually dependent processes of metabolism and transduction cannot be traced back to a known starting point. This acts as an initiator with regard to this chain of processes, but is itself only a link (e.g., a chemical product) of a chain of molecular processes that depends on hereditary and unexplained properties of the organism. I do not need to describe the present level of experimental research on these hereditary properties, but refer the reader to the clear account of Sachs.

Under any circumstances, every initiating process involves some dynamic activity, but the concept of initiation implies not simply this transfer of a certain amount of kinetic energy, but the actual transduction of the potential energy. There does not have to be a definite ratio between the magnitude of the energy actually released and the magnitude of the initiating force. Thus, for instance, the mechanical equivalent of the initiating force, when compared to the force that is released, is infinitesimally small when a large quantity of gunpowder is set off by a spark, but it is also easy to find examples where the opposite is true.

Under known conditions, a whole sequence of processes can be linked to the first initiation process. Thus, the preceding process always triggers the following one. Sometimes, the consecutive links are linked by yet another initiating process, or by simple transferral of energy from one molecule to another. In the course of a reaction, two separate processes may develop, taking place separately but concurrently, on occasion supporting each other or dependent on each other at later stages in the reaction, e.g., when the products of a specific process have a releasing or transferring effect in different directions. Such complex sequences no doubt play a role in the living cell. Here, the structure of the cell may also cause the sequence of processes involving metabolism and transduction that follow an initiating process to become even more complex. For example, it is conceivable that the first step of a sequence of processes begins in the cell sap. These processes develop in a certain direction in the cell sap. When a substance derived from one step of these processes enters the protoplasm, an initiation is stimulated in the protoplasm as well. This initiation process in turn develops in a complex way. The final factor involved may then be the resistance of the cell wall, which varies with growth.

When evaluating externally observable activities of the cell, we should keep in mind that only the resultant of the entire transduction is expressed. Of course, this resultant may be exactly opposite in two types of material if their individual activities are qualitatively but not quantitatively the same. Thus, a body that is pulled by two or more forces can move in one direction or another, depending on the resultant, which varies with the intensity of the forces. For instance, in positive or negative heliotropism or geotropism, processes initiated by light or by gravity may be qualitatively identical in the case that the resultant of the effective dynamic activities points in the positive direction along the abscissa in one material, and in the negative direction in the other. When the initiating force is the same, the simple fact that two different objects curve in opposite directions cannot always help us to determine whether the initiating processes differ only quantitatively, or whether qualitatively different processes were initiated.

Over time, it is possible that the same cell will change its specific properties to such an extent that two unequal phases of development react quite analogously to the same initiating force, as could two different cells. What I just said also provides a criterion for evaluating the activities of the same cell in different stages of development. Among the reasons for focusing on behavior at different times is the fact that geotropic curvature, in some plants, is exactly opposite for the same organ in different phases of development. I would also like to remind the reader that in the same plant the total released activity could be different depending on the intensity of the initiating force. For example, one initiating process is completed following each impulse, while another initiating process takes place only according to the intensity of the mechanical equivalent of the initiating force, and thus rises with the latter.

My remarks, which do not exhaust all possibilities but which were only meant to suggest certain important theoretical aspects, were intended to show in what complex ways the relations between initiating and initiated forces can develop. In concrete cases, researchers need to adapt their questions and experiments to the circumstances.[76] In tissue complexes where there is an association of cells (and because of this association of cells), we have factors of resistance and expansion and general factors that are of importance for the resultant of transduction in the individual cell. But in order to try to explain the total activity of an organ it is necessary to understand that very essential factor, the activity of the cell (considered as an isolated unit). Here, and for a cell that is free in itself, a course of events can obviously be causally explained up to a certain point even if not all the processes are known from the initiating process to the final cell activity that we are trying to explain. We do not have as much reliable data on cell mechanics for processes of expansion and growth as we do for a stimulation movement initiated when certain plants are jolted. For in all other cases there has been no definitive decision as to what extent the pressure of the cell contents and the resistivity of the cell wall are the determining factors. Especially for periodic movements dependent on differences in illumination (or also on temperature fluctuations), the latest state of research is essentially still as I described it on another occasion (Pfeffer, 1875d:113), even though methodological means that might allow a limited decision on the question are available to us. For example, once it is possible to show that in a variation movement the pressure exerted by the cell wall on the contents does not change during such movement, then the cause of the pressure fluctuation is entirely to be found in the cell contents. Among other factors, the monitoring of the volume relationship of protoplasm and cell sap could be decisive in this respect. Whether these and other methods succeed will depend mainly on whether we can find suitable experimental material.

For reasons of probability, which I will not present and expand on here, no one will seriously doubt that the mechanical cause of periodic movements is pressure fluctuations initiated in the cell contents by light or heat. This occurs as well when we are dealing with the expansion of elastic cell walls or with the growth of the cell wall. If this is true for periodic movements of leaf organs, the same cause probably underlies fluctuations of the longitudinal growth, which are determined by the same external conditions. There is no doubt, as Sachs was the first expressly to point out, that the magnitude of the turgor is an essential factor in the extent to which the cell wall increases in area. This obviously also depends on the resistivity of the cell wall, on the material for growth, and on the ease with which new cell wall molecules are intercalated, and other factors as well. Thus there is no question that the so-called grand period of growth is a result of the conditions I have just listed and others as well.

Though no details are now known on the type and the course of processes initiated by light or heat, certain observations do indicate that the process is not always very simple. For example, I have shown for the variation momements, which are induced by a dark period, that the ability of the cells to expand reaches a maximum, and then goes down again quite considerably (Pfeffer, 1875d:93). Somehow, here, one effect must again partially cancel another effect. Either this is inherent in the chronological course of a sequence of processes which follow the initiating process one after another, or else the dark period, right from the start, causes two initiating processes to begin that proceed at different velocities, or there is some other underlying complication. Similarly, the behavior of crocus flowers when the temperature drops suddenly indicates a complex reaction (Pfeffer, 1875d:132). The temperature drop causes a considerable temporary growth spurt. In actual fact, when temperatures are constantly low, growth is slower than at a higher temperature, and obviously growth must be inhibited when the temperature is lower. Obviously resistance,[77] and growth of the cell wall, may be a factor here.

As with the above-mentioned processes, and others similar to them, aftereffect movements that follow a movement induced by a dark period can be explained by these factors (Pfeffer, 1875d:39). The cause of these aftereffects must be sought in the same kind of principles that cause a pendulum to continue to swing, or a function consisting of several variables to give a periodic curve in its graphical form.

23. Heliotropism and Geotropism.

Heliotropism and geotropism show that a unilaterally acting force can cause convex or concave curvature toward the initiating force. Speaking more generally, analogous phenomena will presumably be produced by the corresponding action of a releasing factor. As examples, we might refer to such phenomena as the deviation of roots by moisture and the consequences of their prolonged contact with certain objects.

Now let us examine heliotropism with reference to cell mechanics. First, let me point out that in certain plants the ability of cells to expand decreases with increasing illumination. The most probable reason for this is the decrease of turgor, which we will simply assume here. Then, when the illumination is from one direction, turgor in the cells must increase on the dark side, and a positively heliotropic curvature of the tissue complex is the necessary consequence, if nothing else intervenes. But the positive heliotropism of unicellular organisms cannot be produced in such a way, and thus we must distinguish two types of positive heliotropism (Pfeffer, 1875d:63). The first is heliotropism caused by gradually decreasing cell turgor, which is possible only in cell complexes, and the second is heliotropism of unicellular organisms, which is most probably caused by the influence of light on the cell wall. At the same time it is possible that in one tissue both types of heliotropic curvature are found together, since in addition each individual cell has a tendency to curve for the same reasons as unicellular organisms. Nevertheless, it is, of course, important to distinguish the two types, and we shall speak of the positive heliotropism of tissues and the positive heliotropism of the cell. For the time being both only have one thing in common, the same force triggers the process. We can only discover whether there are other similarities during the triggered processes when we learn more about the processes. As far as negative heliotropic curvature goes, we have nothing to indicate that there is a difference between the heliotropism of tissue complexes and of unicellular organisms. The longitudinal growth of negative heliotropic organs, as far as we know, is slowed down by light in the same way as the growth of positive heliotropic organs (H. Müller, 1876:13). Heliotropism without involvement of growth (Pfeffer, 1875d:63), that is, by a simple expansion (capable of reversibility) of elastic cell walls, is found only in the positive heliotropism of multicellular organs. Since we do know the mechanical causes of heliotropic curvature, it is impossible to say whether such heliotropism is otherwise possible or not.

In unicellular organisms, it is true, --as familiar facts indicate --that an alternately falling and rising turgor cannot be the mechanical cause of heliotropism. Yet, such a fluctuation of the hydrostatic pressure will have a certain influence on the extent of the curving. Assuming the longitudinal halves of the wall of a cylindrical tube are unequally expansible, then, as tugor increases, a straight tube will curve concavely toward the less elastic half of the wall. However, a tube that was previously concavely bent toward the more elastic longitudinal half will obviously, more or less, straighten out its curvature. As the turgor oscillates, for a perfectly elastic wall, and depending on how expandability is distributed, the tube would then describe a plane or a curved surface as it moves back and forth. We must also keep in mind that where the wall is uniformly equally expandable and elastic, the curvature of a tube must be decreased as the hydrostatic pressure inside increases, a process based on the principle of Bourdon's spring manometer.[78] Increased expansion stimulates an increase of growth in growing cell walls, and thus increasing turgor must tend to counteract, to a certain extent, the curvature of a cell. Moreover, the fluctuation of hydrostatic pressure in growing cells will be able to influence how extensive the curvatures and oscillating movements of the cell will be, if conditions for such movements occur in the timewise changes of resistivity in the cell wall and of the cell wall's ability to grow.[79]

The above clearly shows that the decrease and increase of hydrostatic pressure in a cell is a possible factor in positive and negative heliotropism. Fluctuating turgor cannot cause the heliotropic curvature of single cells, since then the decrease of light on all sides would already have to produce a curvature. Curvature, however, and the mode of the curvature, is determined only by uni-directional illumination.

The greater expansion of one longitudinal half of the cell wall, and the correspondingly greater growth, could also be caused by the protoplasm body, which in fact, due to its structure, can exert a certain uni-directional pressure, though of course it is only minimal. It is true that even a slight excess expansion in one longitudinal half of the wall would, when prolonged, be able to cause a cell to curve considerably with time. That is, if by this slight but prolonged excess expansion, the growth of that particular half of the wall would be continuously accelerated somewhat (Pfeffer, 1875d:146). But even a slight counterpressure would then suffice to halt the curving because, as we explained earlier, the protoplasm body, having a high water content, can only exert slight uni-directional pressure.

It is not yet known with how much force the heliotropic curving of single cells takes place, but I have observed that a relatively high pressure was overcome by the positive heliotropic curving of the internodia of a Nitella. The tip of the plant had to push this pressure ahead of it without the curving internodia coming in contact with the solid body. Though I canot give the pressure level, it was too high, at any rate, to ascribe it to a corresponding tendency to expand by the protoplasm body. Thus, I must reject the unlikely assumption that the mechanical cause of heliotropic curving of single cells is an unequal force in the protoplasm body. Then, however, the cause of heliotropic curving must be sought in unequal resistance or growth of the cell wall. Though our considerations and arguments are all related only to positive heliotropism, I do not doubt that they would result in the same conclusions for negative heliotropism --i.e. would see its cause in the cell wall. Of course, we need conclusive experiments for both types of heliotropism of single cells to prove this absolutely.

Now, if positive and negative heliotropism were in fact to depend primarily on molecular processes inside the cell wall, then the same triggering force, acting in the same direction, --i.e., light --must trigger observable activities that act in exactly the opposite direction. Where several variable factors are involved, this is possible (as I pointed out earlier) if the triggered processes are qualitatively identical, but not identical in intensity. And I want to emphasize again that, where we have such movements in opposing directions, it is impossible to decide on the basis of this common phenomenon alone whether they are the resultant of qualitatively or only quantitatively different activities. It is impossible to decide the question on the basis of what we know about heliotropism.

Growth is always a combination of a certain number of factors. We will always be able generally to regard the supply of appropriate material and the use of the material for embedding new cell wall particles as the immediate factors. However, these factors themselves are already the resultant of a whole series of more immediate and more remote conditions. At present we know nothing about how uni-directional illumination acts as a trigger to make the increase in cell wall surface area relatively greater on the illuminated side in one case, and on the shaded side in the other case. Here we could say that light seems to promote the thickening of cell walls (Kraus, 1869-70:232), thus decreasing the cell wall's growth in surface area, because this essentially depends on how much expansion there is. On the other hand, we know that light may promote growth in certain organisms, though how this happens is still unknown.[80] Thus it seems we have two factors acting in opposite ways. If they were initiated in two different cells, in an unequal ratio, they might well cause positive and negative curving. I am not trying to propose a hypothesis, but to demonstrate by an example that it is possible to find ways of discovering the mechanical causes of opposite heliotropism by experimental means. Possibly the uni-directional light will act on the protoplasm body in such a way that the conditions for growth that it provides are unequally distributed over the lighted and shaded side of the cell.[81] Single cells that curve heliotropically without wall growth are unknown. Should they exist, they would be of incalculable value in explaining the causes of heliotropism.

If the positive heliotropism of certain plants is replaced by negative heliotropism at a later stage of development, the effective resultant has been reversed. This is, perhaps, because as the organ continues to develop, the ratio of activities acting positively and negatively changes, or because a new activity is added. The hereditary properties of the plant determine the place where negative heliotropism visibly begins. In accordance with what has been said, this can happen both in the fastest growing and in the slowest growing zone. That is why I cannot agree with H. Müller's (1876:70 and 93) distinction between two types of negative heliotropism, if it is based solely on the occurrence of negative curving in various zones of growth. This is a phenomenon that results, essentially, when two forces simultaneously act on organs in an opposite sense, and their ratio is unequal in different plants at the same stages of development.[82] Of course it is possible that the negative heliotropism of different plants is only the identical common expression for qualitatively unequal processes. However, at this time we know of no data that would lead us to think so.

In order to have a triggering effect, a ray of light must always exert a certain amount of molecular force, though this may be very minimal. We can now ask whether this molecular force is the same when an identical ray of light enters the cell wall and protoplasm (or simply the place where it will act as a triggering mechanism) at the same angle of incidence from the outside or, in changed sequence, from the inside. If the sequence in which light enters is irrelevant, the different effect on the shaded side (which had the same properties before) must be caused by the fact that here we have light of the same quality but of a different intensity. It is inevitable that a ray of light will become less intense after it acts on the illuminated side of a cell, and so the light ray will arrive at the shaded side with less intensity (mechanical equivalent) if the particular structure of the cell or the organ does not cause the rays to become concentrated.[83] If a ray of light should in fact produce unequal effects when it enters the wall from the outside or from the inside, we must also decide right from the start whether the ray of light in both cases had the same intensity when it reached the place where it exerted its effect. Only then could we conclude that the ray of light acts differently if it falls on a cell wall particle from the outside or from the inside. Because a triggering process is involved here, a very minimal difference in the efficiency of the light may already produce an enormous effect.

Heliotropic curving also occurs when a ray of light strikes the anterior and posterior walls of the cell at the same angle of incidence, and this case immediately forces us to ask how and when heliotropism is produced. Another question is how longitudinal growth, or the growth of a cell in general, is changed with the angle of incidence of the effective light. To answer that question, we must first compare processes of growth in a cell in which in one instance rays of light enter parallel to the longitudinal axis, while in another instance they enter at right angles to it, but evenly from all sides. The effect of light rays at different angles of incidence would have to be resolved into these components. But assuming that rays of light that entered parallel or vertical to the longitudinal axis of the cell actually affected the growth of the cell in basically different ways, this in itself would still not explain heliotropism as such. However, the significance of the angle of incidentce of the light rays for the degree of heliotropic curving produced could perhaps be completely clarified.

I felt I needed to make these remarks, since H. Müller's (1876:92) statement that heliotropism involves not so much light difference as light direction does not allow us to ask more precise questions. It is ambiguous because it does not differentiate the incident angle of the light from the direction of the light, with reference to whether the light comes from the outside or inside. Of course, we cannot decide whether in heliotropic curving a light ray acts only because of its intensity, or whether the same ray acts differently if it falls on the outside or inside of a particle of cell wall or protoplasm. Incidentally, the former alternative appears to be likelier in any event. Further discussion on these questions seems superfluous as things are now. I do want to point out that when turgor in the cell decreases, it would only be possible for a light ray to exert an effect other than through its intensity if the necessary triggering process occurred through the effect of light on a non-liquid substance.

If we compare the growth brought about by light entering the cell parallel or vertical to the longitudinal axis of the cell, we must note that displacements, such as are known for protoplasm, and other circumstances as well possibly affect the growth of the cell wall in significant ways. These and other points must also be taken into consideration for heliotropism. It is not even probable that the mean growth of organs that curve heliotropically and of organs that grow straight, under the same intensity of light, is identical.

Until now, as is proper, I have presupposed identical light rays. The effects of light rays of different wave lengths and different vibrational planes present quite another set of questions, which have to be considered both for the ray that is parallel to the cell axis and for that which is vertical to it. As regards polarized light, I am aware of only one experiment by Askenasy (1874:237) according to which the fruitstalks of Pellia and cress plants exhibit positive heliotropism in the same way when the vibrational plane of the light that falls on them from one side is parallel or vertical to the longitudinal axis of the cell. Since it has not been ascertained whether the polarized light was not depolarized again, this experiment is not exactly conclusive, although I have no doubt whatsoever that light of every vibrational plane produces heliotropic curvature. I believe that, for certain reasons which I do not want to elaborate here, it can be expected that a light ray that enters parallel to the longitudinal axis of the cell acts independently of its vibrational plane.

Basically the same types of questions need to be asked regarding the mechanics of geotropic curving as for the heliotropic movements. The existence of positive and negative geotropism in unicellular organisms and similar considerations to those I discussed in connection with heliotropism must lead us to the conclusion that geotropism of the cell is produced, in plant structures inclined toward the vertical, by gravity, which causes triggering processes. These unequally affect the growth of those parts of the cell wall that are oriented upward or downward, though perhaps only very indirectly.[84]

In the tissues with negative geotropic curvature, with or without growth, the expansive force,[85] that is the growth of the cells, increases from the side that is turned downward to the side that is turned upward. With positive geotropism it is the other way round, the topmost cells grow fastest. In the case of horizontal positioning, therefore, a factor must be introduced that differently rearranges the triggering process in all higher cells. But I do not want to form an opinion on whether we need to differentiate between the geotropism of the cell and the geotropism of the tissues, as in positive heliotropism. The data available at this time do not absolutely require such a differentiation.

Compared to the enormous expansive force that some structures may develop in negative geotropic curvature, the energy of gravity necessary for the triggering process must be infinitesimal (Pfeffer, 1875d:146). It is still impossible to say anything definite about how this triggering process is brought about, whether the triggering process is qualitatively identical in positive and negative geotropism, or whether two triggering processes are involved, or similar problems.[86] Of course, the idea suggests itself that the slight excess pressure is a triggering factor that is exerted in cells and in tissues in deeper layers by the weight of the liquid column above them. But perhaps the deflection of roots by moisture is a phenomenon that is based on similar principles as regards the triggering process (Sachs, 1872:219). I shall not go into detail as to how this is to be understood, but I did want to point out the possible connections and draw attention to the fact that, in any event, it is necessary first to study how negative geotropic organs and unicellular organisms react under those moisture conditions that cause the positive geotropic roots to deflect concavely toward the moister side.

I will not go into a further analysis of the question whether direction or intensity of gravitation is important for geotropism. What I have said about light contains the most basic points that would need to be followed up concerning gravitation. Incidentally, it is obvious that gravitation acts upon two points, whose vertical distance from each other is the height of one cell, in a manner so nearly equal that the cause of the triggering process must be sought not in this difference but in other processes related to gravity.

24. Some Growth and Formation Processes.

The purpose of this treatise is not to deal with everything that can at present be said and conjectured about cell mechanics, and yet I believe I need to touch on a few points related to growth processes that were not easy to incorporate in the preceding sections.

If a certain excess osmotic pressure is present in the cell, the plasma membrane that forms the outer covering of the protoplasm body must necessarily be pressed close to the cell wall. The properties of this plasma membrane and of the protoplasm body itself make it unavoidable that even the minutest pit areas in the cell wall will be lined with it.[87] Now osmotic pressure seeks to expand the surface area of the cell wall, and also acts so as to compress it transversely when the cell wall or certain layers of it rest on a support. For when this is the case the cell wall, or rather an inner part of it, is pressed together in a way similar to a mass, capable of swelling, wrapped in a cloth and put in a press. This mass extrudes water until the external pressure is no longer capable of pressing out the imbibed water, which is retained by a greater force.

Mostly the cell wall becomes only very slightly thinner in this way. But the decrease in thickness must become considerable when the cell wall --though very much increased in volume by swelling --only retains the liquid it has imbibed with such [*small*] force that a given osmotic pressure can press out quite a large quantity. This is evidently so in certain algae filaments whose cell wall may swell to double its original thickness in sucrose solution, or after the cell is killed, i.e., generally after the turgor is abolished. Meanwhile its length does not change significantly (Nägeli and Schwendener, 1867:406). Here, the external resistance and slightly expandable layers of the cell wall form the support against which the inner swelling layers of the cell wall can be pressed by the plasma membrane, that is the osmotic pressure. No doubt numerous analogous phenomena can be found when we focus on plants in which the cell wall is transformed into masses that are capable of swelling. But even where there are no perceptible diameter changes in the cell wall, the compression I referred to must not be ignored, since both the slightly diminished water content and the pressure itself might be a factor in growth and in other processes. The mechanical causes of new formation and growth of cell wall can as yet not be explained in a generally satisfactory way. When a new cell wall forms around a protoplasm body, the material needed to form the cell wall, which comes through the plasma membrane in a dissolved form, must be precipitated in insoluble form. My conjecture is that the cause of the decomposition of the solution of membrane-former and the formation of the cell wall must be sought in the contact with an aqueous medium, with air or with any other medium. This is so not only where the cell wall forms around a protoplasm body lying freely exposed in water, but also where it is formed when there is cell division with two previously separated protoplasm bodies.[88] There are other conceivable reasons why the membrane-former could be precipitated out of the diosmosing solution. I shall not discuss them, since there are no definite facts available on the mechanical cause of cell wall formation.

Regarding the causes of growth of the cell wall in surface area and thickness, no basically new points of view have been added to the ones which Nägeli developed in his incomparably perceptive studies and conclusions on the growth of starch grains, by means of which studies he also made it clear how the cell wall grows (Nägeli, 1858:328,370). He also clearly recognized that one cause of growth was negative tension of the cell wall, but of course he did not believe that the expansion of the cell wall by turgor had great significance for surface growth of the cell wall, as Sachs (1873:699) correctly felt it did. Though it was important simply to point out that turgor plays an important role because it expands the cell wall by growth in surface, a basically new cause of growth has not been introduced by this concept. Nor did Traube introduce any new concepts, though he was able to demonstrate a few of the causes for growth and growth processes, in obvious ways, when he made his important discovery of the precipitation membranes. However, it does not follow as a matter of course from these growth phenomena in precipitation membranes that the cell wall, too, grows only for the same reasons. For example, though the surface growth of the cell walls is very much slowed down when turgor decreases, it is still uncertain even today whether cell walls do not grow into the surface, in certain cases, without passive expansion, perhaps as other molecular forces active in the cell wall cause new cell wall particles to be incorporated. Such a growth process is necessary at any rate to make the cell wall grow thicker, for transverse compression by the variable osmotic pressure cannot be the cause of increasing thickness, though thickness may be somewhat affected by this compression.

I do not intend to discuss the causes for the growth of the cell wall any further, by the way, and only wrote the preceding to point out that, as Nägeli made clear almost 20 years ago, a whole series of factors are involved in causing growth. Among these factors Nägeli also emphasizes those which Traube considers the sole causes of growth --the presence of the membrane-formers and negative tension. A calm examination of the known facts will easily show that these alone, or in combination, do not suffice to explain the specific growth of cell walls.[89]

The formation and distribution of the substances that mediate growth are among the factors that will be taken into consideration for the growth of the cell wall and of the cell as a whole. Because of the special organization of such processes there is a possibility that transduction will be influenced by metabolism in many ways, and vice-versa. Yet I do not wish to portray such reciprocal relationships, but merely refer to a few possible causes for locomotion in the protoplasm body. It seems more probable than improbable that such locomotion (e.g., the accumulation of protoplasm in certain locations) is a factor in growth and triggering processes. As matters now stand, I can only draw the reader's attention to a few causes that have not, or not sufficiently, been taken into account, and that may be important in observable formation processes in the protoplasm.

At this time we do not know a great deal about how external influences affect formation in the protoplasm. Light can cause certain movement processes, higher and lower temperatures and certain chemical reagents can cause changes in form. Also, it seems probable that other factors, like gravity,[90] electricity, etc., have an influence in causing movements in the protoplasm in a specific direction, particularly when they act uni-directionally. For example, if the cell sap has a lower specific gravity than the protoplasm, the vacuole must exert a corresponding vertical upward pressure on it, and it will depend on how large the pressure is in relation to opposing resistance forces whether a visible effect is produced in this way. Similarly, of course, gravity must tend to separate solid bodies according to their specific gravity (Nägeli and Schwendener, 1867:381).

In addition to these and other causes for movement, I believe I need to draw special attention to processes that might well be produced by chemical actions.

If at any point in a solution a solute is decomposed, there is sure to be a diffusion movement. Mass motion too, however, can occur when the undecomposed solute and the products of its decomposition (as is usually the case) do not have the same volume. To make the matter as simple as possible, we could imagine a case where water is split off or transformed into a chemical compound, with accompanying increase in density. Diffusion flow and mass flow continue as long as they are produced by chemical decomposition, and may become continuous if chemical decomposition is continuous. The latter is possible, even when there is a small quantity of material, if the decomposing force is always active at one location while at other locations further away from it the substance that is involved is somehow being regenerated. We have to admit that such a thing can be possible in a living cell. Substances that dissociate in the light but are regenerated again in the dark demonstrate how this can even be achieved in a somewhat simple way. Now when such a continuous process is taking place, the mass flow that is produced, e.g., by volume increase, is supported by volume decrease as regeneration occurs at other locations. Meanwhile the diosmotic flows, which in themselves cause no volume change, continue to carry the undecomposed solute to the place where decomposition is occurring, and to take away the products of decomposition from it to the places where they are regenerated. Should such flows be occurring in the protoplasm, then it is clear that undissolved solutes can probably be brought into specific configurations if the effect of these flows --doubtlessly never more than minimal --is not overcome by other forces. I dare not even conjecture whether the kind of configurations that are found, e.g., when cells divide

(Strasburger, 1875), are actually produced by this type or similar types of chemical actions, or whether they are caused by the organization and the specific molecular actions related to it, or in some other way.

When we have a protoplasm body, which forms a spherical covering shell, for example, around the cell sap, flows of the type indicated here will also be able to influence the form of the covering. Where the volume increases, fluid flows away in all directions and, accordingly, produces an impact on the plasma membrane that continues with the decomposition processes. The plasma membrane, as I showed earlier, may bulge, even when there is a very minimal force towards the cell sap, if the hydrostatic pressure on both sides of the plasma membrane remains constant. On the other hand, there is a small excess pressure towards the plasma membrane from the cell sap side when a density increase results from a chemical process in the protoplasm and causes a water flow. Each of these minimal effects, or both together, could gradually even produce important morphological changes in the protoplasm, if it is only a matter of overcoming minor resistance forces. Incidentally, it must be remembered that the water flow which hits the plasma membrane may cause a certain amount of water to be pressed through the plasma membrane and may thus at the same time become the cause of local flows in the cell sap, proceeding in specific directions.

Obviously, what I just said about the effect of flows on the form of the protoplasm body is also true for flows produced in any other way. Here I will simply point out how such flows may come about osmotically, if the osmotically effective substances are not uniformly distributed in the solution. They could occur as a result of local chemical reactions, since it takes time before a completely homogeneous distribution has been achieved by diffusion.

To visualize the situation, let us imagine a hollow glass cylinder that is closed at both open ends by membranes having the same properties and that is placed vertically in water. Let us imagine a more concentrated solution of a non-diosmosing substance in the lower part of the glass cylinder, while a more dilute solution adjoins the upper membrane.[91] The more concentrated solution, by itself, would cause a higher osmotic pressure than the more dilute one. In the cell itself, however, the final pressure must lie between these two pressure readings. This will obviously be reached when, through both membranes together, the same amount of water is being filtered by pressure as is being brought in by osmotic force. The osmotically produced inflow in our apparatus causes a larger amount of water to pass through the lower membrane per unit time than through the upper one. While through both membranes, equal amounts of water (in terms of the surface area) are filtered under a pressure that is approximately the same throughout the whole cell. Thus, a water flow must pass through the cell from the lower to upper membrane as long as solutions with unequal osmotic effects adjoin both membranes. In short, we are dealing with a circulating water flow. Now let us suppose that in the protoplasm body, which forms a spherical covering shell, the solution does not have the same osmotic effectiveness in every place. Here, too, a circulating water flow that goes through the cell sap and protoplasm can be set-up. It can affect the form of the protoplasm body in the manner previously described.

We have discussed chemical and osmotic processes as possible causes for movement and morphological change in the protoplasm (and cell sap). No matter how the chemical reactions are produced, whether their origin is autonomous or induced, the further results of their effects may turn out to be identical. I shall not discuss how these may develop in very complex ways inside the living cell. I would simply add that the location of the decomposition could also be constantly shifted about. We should keep this possibility in mind, especially as regards autonomous triggering processes. Moreover, combinations with other factors are possible, and I remind the reader that the relative volume of the protoplasm and cell sap must change if the osmotic effect in one of these bodies varies. For example, if the cell sap were to increase in volume, and the protoplasm body, which becomes shrunk, sets up unequal resistance at various locations, a mass flow in the protoplasm could also be brought about by this. Repeated oscillation of the pressure, combined with altered resistivity in the protoplasm, would even cause complex flows.

What we now know about movements in the protoplasm due to a change in light (Sachs, 1874:721) does not allow us to conclude by what specific processes the locomotion of protoplasm is brought about. Nothing certain is known yet about the immediate mechanical cause of the rotation and circulation movements of the protoplasm. I shall not give an account of the present state of affairs. I merely wished to draw attention to a few points that should be kept in mind when formulating questions for experimenal study.

25. Upward Transcellular Flow of Water.

There has probably never been any question that the so-called root force is brought about by the osmotic activity of cells, and the fundamental question is how a single cell is capable of extruding water in one direction. In fact, we know of unicellular organisms that secrete drops of water in certain places (Sachs, 1874:659). However such water secretion cannot, as was always assumed until now, be ascribed to specific properties of the cell wall, but must be sought in the plasma membrane or in the cell contents, which puts the cell mechanics of such a process in a very different light.

In order to understand the principle of this matter, we only need to consider the plasma membrane that is pressed to the cell wall. Let us assume that none of the osmotically effective substances is diosmosing out of the cell, which must first be imagined freely immersed in water.

Assume water is to be pressed out, that is, in other words, a transcellular water flow is to occur. Then the quotient obtained from dividing the inflow (caused by osmosis) by the outflow of water (produced by [*hydrostatic*] pressure) cannot equal one at every part of the surface area of the plasma membrane. It must have a larger value at least at one part of the surface area and a smaller value at another. However, the pressure, i.e., the ratio between inflow and outflow, is independent of the membrane thickness, as I showed earlier. It follows that at least the uneven thickness of an otherwise homogeneous plasma membrane cannot become the cause of a unilateral continuous water flow. Thus filtration resistance cannot be the cause either, in as much as it only depends on membrane thickness.

A transcellular water flow can only be produced under two conditions --if there is a qualitative difference in the plasma membrane, or if the membrane is homogeneous but the same osmotic effect does not act at all its parts.

Two qualitatively different precipitation membranes can, as a matter of fact, each have a different osmotic pressure with the same solution. Let us assume the pressure for membrane A is higher than for membrane B. Now if one half of a cell was formed from membrane A and the other half from membrane B, the final pressure would lie between the pressures resulting from cells consisting only of A or only of B, for the same solution. But then, in a state of equilibrium, the water outflow produced by pressure exceeds the osmotic water inflow for membrane B, while for A the opposite is true. Consequently, a water flow moves through the cell going from A to B. Its magnitude is determined by the difference between inflow and outflow in each of the two membranes.

However, if the membrane is homogeneous, while in different zones of the cell there are solutions whose concentrations are different, or which have different osmotic effects, the ratio between inflow and outflow for two given sectors of the membrane surface is unequal. A transcellular water flow must pass from the more effective to the less effective solution, as I explained earlier (p. 222). The same result is achieved if the solution in the cell is quite homogeneous, but the cell wall is in part saturated with a solution, and in part saturated with water (or with a less effective solution). Where the solution is in contact with the outside of the plasma membrane, the resulting osmotic pressure is reduced by the osmotic effect produced by this solution across the plasma membrane.

It would not be difficult to construct an apparatus to demonstrate the extrusion of water under the conditions I have described. To do this, one could close a glass tube at both ends with porcelain plates or porcelain cells at whose surface would be either the same, or different, precipitation membranes. If the lower end of the apparatus was placed in a water bath, while the upper end projected into a vertical tube [*an osmometer*], the pressure that would be produced in this way could then be measured. Membranes of copper ferrocyanide and of prussiate of iron would be useful as the different membranes. They each show a different pressure with the same sucrose solution. When homogeneous membranes are used, this involves a solution concentration in the cell that decreases from the bottom to the top. In the other case [*different membranes*] the inside contains a homogeneous solution, and the vertical tube would have to be filled with a more dilute solution of the substance. The pressure measured by means of the vertical tube, when it corresponds to the equilibrium state, indicates by how much the pressure produced by the lower membrane exceeds that produced by the upper membrane. This is so whether the cause of the uni-directional outflow of the water is the qualitative difference in the membranes or, independently of the foregoing, whether it is the osmotic effect of the solution (or solutions) adjoining the membrane.

This uni-directional excess pressure would not be totally measured in a plant cell with the upper end of the cell fitted into a vertical tube. This is because the pressure produced in the tube will extrude a certain quantity of water through the cell wall (which is capable of imbibition) that is located between the place of attachment of the vertical tube and the plasma membrane. Conditions turn out to be even less favorable in tissue complexes, whose hollow spaces as well as inactive and less active tissues provide filtration paths. The root pressure measured in a stump proves only that in single cells water must be extruded in one direction with greater force than the pressure in the manometer shows.

If a completely turgescent cell is surrounded by air, a continuous circulating water flow can move through the cell contents and within the cell wall when water is extruded across the plasma membrane in one direction and evaporation is excluded. The water that has been extruded is carried off in the cell wall and there again absorbed by the plasma membrane, at the place where the osmotic water inflow exceeds the water outflow. But when water is supplied to this place in another way, water droplets can obviously be secreted at the other end of the cell when the osmotic requirement can be more easily met by the water that has been supplied.

The osmotic effect of non-diosmosing substances inside the cell is regulated only by the plasma membrane. When there is no cause for water to be extruded through it in one direction, the properties of the cell wall cannot cause such a process to take place, that is, obviously, unless the cell wall is saturated with osmotically effective substances. It is true that the cell wall will affect how extensive the process will be and how it will proceed --the cell wall's permeability is a factor in water entering or being secreted. For example, if a cuticularized cell wall is only slightly permeable to water, then water will be exchanged (or extruded) essentially only at the places that are not cuticularized.

If osmotic pressure in the cell fluctuates, the pressure produced by uni-directional secretion of water (all other things being equal) changes in a way that can be easily inferred. When tissue tension increases in tissues, cells can be compressed; then the water that has been secreted flows preferentially in the direction of least resistance. The thickness of the plasma membrane and the permeability of the cell wall are then also important in regulating this water flow. However, even such a periodic compression and re-expansion cannot bring about a continuous water flow, since at the place on which the larger part of the water is secreted, the larger part of it is again reabsorbed when compression decreases. Of course, if unequal resistance forces hinder the water being carried off in the two opposite directions, a definite amount of water would be carried in one direction. Yet, even if this were to happen, the kind of continuous water flow that appears in bleeding plants cannot be produced by such periodic oscillations. This can be immediately understood without special explanations. It is also clear that such oscillations of the tissue tension are indeed able to influence how extensive the water flow is.

If it is true that osmotically effective substances do diosmose, contrary to our assumption up till now, the general points of view developed here are still valid. I do not feel it is necessary to explain in detail any potential complications, since under any given conditions these can easily be inferred. But note how, for instance, in tissue the solutions that saturate the cell walls can become more effective osmotically, and thus uni-directional secretion of water from the cells can be intensified.

The credit for really attempting to explain the uni-directional secretion of water from cells goes to Hofmeister, as we know. His train of thought was well chosen, though the explanation he gave and other explanations based on his were, in fact, inadequate as long as the significance of the plasma membrane was not recognized. I do not feel I need to go into detailed discussion regarding certain errors in the interpretation of those experiments that were done with apparatus constructed by Hofmeister or similar apparatus (see Sachs, 1865:207). Suffice it to say that the significance of filtration resistance was incorrectly understood. It is also generally overlooked that filtration resistance --in as much as it depends on membrane thickness or on the quality of the membrane --is basically not always equally important for osmotic effectiveness. It was also overlooked that in the vertical tube of the apparatus, because of the diosmosis of the substance inside the cell, a relatively more concentrated solution accumulated than in the rather large quantity of water in which the cell was immersed.

If uni-directional secretion of water from the cell is due to qualitatively unequal properties of membrane sections, such a water flow can continue indefinitely without a change in the system, provided none of the solute exosmoses. But this situation is similar, in principle, to that in any cell whose osmotic pressure remains constant. That is, the number of water particles that move out of and into the cell in unit time is the same. Only in our case, outflow and inflow are unevenly distributed over the surface sections of the plasma membrane. Obviously, by means of friction, etc., work (force) is also transformed into heat here. But heat is also brought from the outside and keeps the kinetic energy of the osmotically effective molecule, and the osmotic effect that is dependent on it, unchanged if the temperature remains constant.[92]

But if the transcellular water flow is determined by an unequal distribution of osmotically effective substances in the cell, it can only continue until a homogeneous solution has been produced in the cell by diffusion or any other type of mixing. Sinmilarly, the water flow produced must decrease gradually when the surface sections of the membrane are in contact, at the outside, with an osmotically effective solution. Such a water flow would stop completely only after infinite time, if the extruded liquid served only to dilute the outside solution and if the latter would always come in contact only with a particular section of the membrane surface. But if this water flow is to continue unchanged, other processes must continue in order to obtain a solution of equal osmotic effect outside the cell. Also, the water flow, which is created inside the cell by the unequal distribution of the osmotically effective substances, can only continue if a homogeneous mixture in the protoplasm is prevented by certain factors. We must admit that there is a possibility of such processes in the living cell, but at the same time it is also apparent that the elimination of factors that prevent mixing might help explain the cause of water extrusion from cells. Of course, the water flow would have to decrease and stop if it did not depend on a lack of homogeneity in the suface sections that constitute the membrane.

Experimental studies to date have almost exclusively been on the so-called root force. The pressure that was measured is of course only a resultant of the force of buoyancy, the quantity of water filtered into the interior of the organ, and of the filtration toward the outside that begins when a pressure is created inside and whose magnitude is dependent on various circumstances. I shall not describe these in detail, or the factors generally involved in root force, but will merely draw attention to a few individual points. First of all, it should be noted that the cell walls that form an organ's periphery are not homogeneously permeable, or possibly are impermeable to water, as cuticularized or suberized cell walls show. But where the nature of the peripheral cell walls admits water exchange, it will be important whether all the peripheral cells secrete water toward the inside. Assuming all peripheral cells do so with equal force, in an organ whose periphery is formed by cells that connect without intercellular spaces, then water can be extruded only through the cells themselves and through their side walls. The actual pressure results from the amount filtered, on the one hand, and the lifting power of the individual peripheral cells (assumed to be equal). Conditions are more complicated as soon as all cells, in those zones of the organ in which the covering cell wall permits passage of water, are not equally effective. In all cases, however, the degree of permeability of the peripheral cell wall, or of its outermost layer, may also play a role. It is easy to understand why, under given conditions, the pressure for a particular degree of permeability must become a maximum. Incidentally, I shall not show how the pressure is also dependent on the form of the paths available to the ascending and descending water flow inside organs.

Hofmeister and other authors believe root force resides in the root tips, and in fact cut-off roots can secrete water by themselves, but such experiments do not show whether cells of the stalk and of other organs are not equally active. Though the lower pressure indicated by manometers applied higher on the stalk makes one suspect that it is primarily the root that develops the motive force, it is possible nevertheless that other cells are active to some degree. In fact, according to Sachs (1874:660), pieces of young grass stalks secrete water at one cut-off end if the other cut surface can absorb water. Of course, in nectaries and in some other organs, there is also secretion of water without root force.

It is easy to see that if we had a chain of cells formed out of artificial apparatus, in which water would be allowed to pass only through the separating transverse walls,[93] water would be extruded uni-directionally whether only the lower, upper, or middle cell, or whether all cells were active in this respect. Similarly now, a uni-directional water flow in tissues may owe its motive force to cells located near the root tip or elsewhere. To be sure, it is most advantageous, for organs in which inactive tissues and spaces are present as well, if the peripheral cells filter water inward wherever the organ is externally closed off by cell wall that is permeable to water. Under certain circumstances it is even possible that unilateral filtration of water from internal tissue cells does not produce a water outflow in the transverse section of the organ at all.

After what has been said, there are only a few remining possible causes for uni-directional water flow out of cells. It is impossible to say with certainty which of the alternative possibilities is determining in the plant; that question can only be decided by specific investigations. Wherever the imbibed solution is not homogeneously distributed in the cell wall, i.e., produces an unequal osmotic effect in its contact with the plasma membrane, a unidirectional water flow must necessarily be developed in this cell even though the cell wall almost always only imbibes solutions of low concentration and, though only the difference in the osmotic effect of these solutions is relevant, this factor must by no means be underestimated. For example, a 1% solution of potassium nitrate, whose specific gravity is almost 1.006, raises a mercury column 175 cm with a copper ferrocyanide membrane, and thus produces a higher effect than has so far been measured for the root force.[94] Obviously, even pressure produced by a 0.1% potassium nitrate solution would still be quite respectable, and any minor difference in the osmotic effect of the outside solution in contact with a cell must give rise to a specific unidirectional water outflow from it.

The conditions for the development of a root force are provided, for example, if a root is immersed in a solution, which is less effective osmotically than the liquid imbibed by the cell walls that are lateral and oriented towards the inside. By metabolic processes, and by exosmosis of substances from inner cells, these conditions can be continuously maintained, not only in the peripheral cells of the root but in the inner tissue cells as well. Specific investigations will have to show to what extent uni-directional water outflow can be produced in this way. Knowing the specific gravity of the liquids and the ash constituents, obviously, is not sufficient when we are interested in the osmotic effect. So, I will not draw conclusions from a comparison of the composition of a nutrient solution and of the liquid that flows out of a bleeding plant. Incidentally, osmotic inequality in the cell wall can be intensified or produced by the streaming of the sap outflow. In this respect, the same plant would not have to react in quite the same way if it bleeds from a cut or if the liquid inside is only under high tension.

In order to cause a drop of water to flow out of individual cells, apparently so minimal a difference in the osmotic effect of the cell wall liquid is required that it could also exist in the cell wall of a unicellular plant. Nectar secretions and similar secretions of liquid may well come about in the manner considered here. Once an osmotically effective solution is present outside the cell, it must act as described above. One of a number of possible factors involved in maintaining constant the conditions that are necessary for water outflow to occur is that perhaps a small quantity of colloids is dissolved in the nectar, and that the cell wall barely, or not at all, imbibes these. The initial cause for such a uni-directional water outflow is probably provided by the accumulation, in some way, of an osmotically effective solution in an appropriate location. Either the cell contents give off a substance to the outside for that purpose, or a soluble substance forms by the transformation of the cell wall, or is brought there in some other way.

Still, the water flow caused by the osmotic effect of the solution in the cell wall is not enough to produce root force. Under certain circumstances the water flow might turn out to be in a negative direction. Moreover, as I showed earlier, differences in the properties of the plasma membrane and, where the membrane is homogeneous, unequal distribution of the osmotic effect of the cell contents may cause localized outflow of water. At this time it is impossible to say whether and to what extent one of these factors or both of them are significant for root force, and for the extrusion of water in general. A limited variability of the osmotic effect at different parts of the same protoplasm body is probably likely if metabolic processes take place in such a way that equalization by diffusion is never achieved. Incidentally, it must be apparent according to what I said above that not every outflow of water from a cell is produced in the same way.

Another matter that must still be determined experimentally is through which mechanical processes the periodicity of sap outflow is brought about. This is where we need to consider the fluctuation of tissue tension which, by means of altered pressure may, but need not, cause the absolute amount of the outflow to increase or decrease to a certain extent. Since the data we have are uncertain, it is impossible to say with certainty if the height of a column of mercury raised by root force also undergoes daily or other periodic fluctuations and, if this is the case, what relationship there is between the pressure fluctuations and the amount of outflow. For of course this outflow depends not only on the driving force, but also on the filtration resistance of the plasma membrane and other resistances. Thus, it is impossible to say a priori whether greater pressure is related to a lesser or a greater amount of outflow. Incidentally, a careful study of pressure and the amount of outflow produced by root force, keeping in mind other periodic processes in the materials studied, may provide the means of understanding the cell mechanics that underlies root force.

The phenomenon of bleeding as it occurs in topped plants is always a hereditary process. The amount of bleeding could, at the most, be slightly altered by turning the plant upside down.[95] For the individual free cell, and also for individual active cells in tissues, experiments are needed to decide whether the process is hereditary or induced. The latter possibility cannot simply be dismissed out of hand as long as it is not known whether gravity is related to the direction of the water outflow, which may again be true only for certain plant materials. But should there be a connection, then experimental studies might enable us to find important clues in order to understand the release process that produces geotropic curvature.

26. Summary of Some Results.

I emphasize only those results from the physiological part that form the basis for all the conclusions reached there.

Physical part

The particles of a substance probably cannot pass through Traube's precipitation membranes without entering the field of molecular forces extending from the membrane particles (molecular osmosis).

The relative molecular size of solutes cannot be determined by diosmosis as a matter of course.

If the effective substance does not diosmose, the maximal osmotic pressure will be brought about in a given membrane. Accordingly, if membrane particles continue to move closer together, all other things being equal, there will be no rise in pressure.

The pressure is independent of the thickness of the membrane. Obviously, the amount of water flow directed into a cell decreases with membrane thickness.

Osmotic driving force depends on molecular forces that interact between membrane particles, water and solute. A zone with an altered composition, the diffusion zone, is formed by these molecular forces at the surface of the membrane.

Since, in addition to the composition of the diffusion zone, attraction between the particles of solute and of water determines the osmotic driving force, the pressure for a substance that diffuses rapidly will generally be higher than for a substance that diffuses slowly, if no exosmosis takes place. There can, of course, be no simple relationship between the pressure and the diffusion constant, because different substances form diffusion zones of unequal composition.

Accordingly, crystalloids in Traube's precipitation membranes produce a disproportionately higher pressure than do colloids. On the other hand, in parchment paper, animal membrane, etc., the effect of the colloids can occasionally even exceed that of crystalloids. This is because the effect of these easily diosmosing substances [*crystalloids*] lags much further behind the maximal pressure than the effect of the crystalloids [*colloids*], which diosmose with more difficulty.

Osmotic pressure increases with the concentration of the solution in a ratio particular to each solute and each membrane. In the same membrane, unless the effective substance diosmoses, the osmotically produced water inflow and the pressure increase in approximately the same proportion, that is the amount of water filtered is proportional to the pressure. This also indicates that the composition of the diffusion zone is not significantly influenced by a uni-directional water flow.

Fluctuations in temperature, which expand the spaces between the membrane particles, will not affect the pressure as long as the effective substance does not diosmose. Generally, however, pressure fluctuations will be produced by outside interference when, by modifications in the membrane or in the cell contents, the composition of the diffusion zone or the molecular interaction between the water and the solute is altered.

Physiological part

Whether a solute is absorbed or not absorbed into the protoplasm is determined by a peripheral layer of the protoplasm, the plasma membrane, which is doubtlessly formed wherever protoplasm contacts another aqueous solution.

A substance that diosmoses through the plasma membrane must diffuse in the protoplasm, or in the cell sap, unless special processes, such as chemical binding, fix the substance that has entered at particular locations.

The high pressure in plant cells is caused by the osmotic effect of solutes contained in the plasma membrane, in which, as in certain artificial precipitation membranes, crystalloid substances are the most effective.

Since protoplasm is also delimited from the cell sap by a plasma membrane, the cell osmotically resembles a system formed out of two cells of different sizes fitted inside each other.

Notes: Physiological Part

1. Then by protoplasm I mean the entire protoplasm body, as is generally common usage now. By cell sap I mean the fluid found in the spaces enclosed by the protoplasm body (as Mohl probably also used the term), see Sachs (1874:2).

2. This was my earlier intention, and so, in a preliminary paper (Pfeffer, 1875c) I called our present plasma membrane a primordial utricle. Mohl (1855:701) notes that the primoridal utricle (i.e., the hyaloplasm) is probably covered by an additional membrane-like layer of the same type, our author suspects, as that which encloses the chlorophyll grains. The term "pellicle" which he proposed as a name for the plasma membrane did not seem an appropriate one to adopt. Incidentally, Mohl (1855) protests against such a limitation of the term "primordial utricle" as I had intended earlier.

3. Mohl (1844:294) and Nägeli (1844:95) were already aware of the indefinite delineation of the hyaloplasm against the granular plasma.

4. In the literature, deBary (1864:51) distinguishes a peripheral layer and a covering. I leave open the question whether this covering is not perhaps something different than the actual diosmotically decisive layer.

5. Mohl (1844b, 1846 and 1855) at least later called the hyaloplasm the primordial utricle. However, in his first paper, he also gave this name to the entire wall protoplasm, which forms a thin layer.

6. Hartig's (1843) first detailed account is not available to me, Hartig (1844:8 and 1858:1, 23) also provide detailed accounts.

7. Strasburger's book (1876b), which did not appear until after the completion of this manuscript, gave me no reason for changing my views on any point. Strasburger has not taken the diosmotic conditions into account at all, and the arguments he puts forward against the existence of a membrane do not need to be specifically refuted after what I have already explained and plan to explain in what follows. I shall make a few comments on

Strasburger's views when appropriate.

8. When I refer simply to Vaucheria from now on, I mean this species.

9. See Nägeli (1844:91 and 1855:9). Also see Hofmeister (1867:76) and Hanstein (1872). See also Strasburger (1876b:26). I was already aware of the fact that Strasburger reported here about Vaucheria, and I could add further facts on the subject. But while Strasburger concluded that a body which forms cellulose could not be produced from granular plasma alone, I felt his conclusion was premature; I still think so.

10. The diosmotic properties of protoplasm masses immersed in oil could perhaps be determined by using suitable substances that are soluble in oil and in water. For protoplasm masses surrounded by air, inferences can perhaps be drawn from their diosmotic behavior toward gases.

11. See this treatise, p. 39.

12. Link (1833), also see Mousson, (1871 and 1871-74:258).

13. Strasburger (1876b:38) thinks vacuoles are probably usually bounded by such a "physical surface film" alone. But this cannot be true because of the diosmotic behavior.

14. See, for example, E. duBois-Reymond (1858 and 1870) and Quincke (1870). Cintesoli (1876) proposes different views.

15. For example, it is still an open question whether crystals and other substances are covered with membrane-like plasma mass already inside the cell, or whether this layer is only created when the cell is killed.

16. Thus, the cell wall is also formed as a precipitation membrane, because the constituent particles can only penetrate through the peripheral layers of the protoplasm in dissolved form. I do not know how Strasburger intends to delimit the term "precipitation membrane" if the cell wall is not regarded as one, as he

expressly says (1876b:37). But incorrect under any circumstances is Strasburger's opinion (1876b:37) that the existence of certain structural relationships in the surface layer of some objects *[plant cells]* proves that the latter cannot be a precipitation membrane.

17. Traube (1867:129) -- I (1873:134) have on an earlier occasion already pointed out that the plasma membrane of the primordial utricle is possibly formed in an analogous fashion.

18. In certain solutions of proteins, carbonic acid produces a precipitate. See, for example, Heynsius (1874:544). For obvious reasons, we might also think that oxygen would have an effect.

19. The destruction of the membrane-forming material in red beets can be accelerated when individual cells in sections die off, and thus their acidic cell contents escape into the outside medium. Incidentally, the behavior of the isolated vacuole indicates that even without external addition of acid the solution of the membrane-former is destroyed over the course of time.

20. See Graham (1862:30). I do not feel Graham's statement that the diffusibility is not at all decreased by the influence of the colloids is quite correct. For according to Marignac (1874:561) the diffusion velocity of each individual solute is decreased in solution mixtures, yet this decrease is only slight for the most diffusible solute.

21. Among them is the considerable pressure produced in plant cells even when the concentration of the effective solution is low. In this respect, too, the plasma membrane acts in a similar way as a precipitation membrane with narrow interstices. More about this in later sections.

22. Those who wish to assume that plasma membrane does form around vacuoles, but does not exist around living protoplasm, must explain the identical behavior, diosmotically and otherwise, i.e. a fact, before their assumption can be justified. Such an assumption is not even supported by visual evidence. Obviously, in these questions about the physical constitution of the surface boundary of protoplasm, we are not interested in the

particular functions carried out during life in the plasma membrane and by means of it.

23. Acids and ammonia penetrate the cuticle layer of the filament hairs of Tradescantia only extremely slowly, but pass quickly, however, through the transverse wall that separates the cells. Therefore, starting with a cell that has been cut apart, these substances diffuse out and the color reaction is found in adjacent cells with correspondingly different degrees of intensity.

24. The effect of ammonia was already observed by deVries (1871a:8 of the reprint, and 1871b:36 of the reprint).

25. See Hofmeister (1867:53), where the same information is given about potassium hydroxide solution.

26. Obviously substances can act on protoplasm only if they can penetrate the plasma membrane. Experiment must decide if this is the explanation for the lack of response to specific poisons, such as veratrine, or whether these substances do not harm the protoplasm. Kühne (1864:100) found the protoplasm in filament hairs of Tradescantia still streaming after the material had been immersed in an aqueous solution of veratrine for 17 hours. On the other hand, the same author (1864:86) claims that myxomycetes die in aqueous veratrine solution. In microspic sections of red beet, which were placed in a saturated (approx. 5%) solution of morphine acetate, I found individual cells preserved with their cell sap still normally pigmented even after 10 days. I have not tried to learn more about the type of effect produced by this and other poisons.

27. These plasmodia had been immersed in water once before to prove beyond doubt that the plasma membrane exists. The plasma membrane can be present in contractile vacuoles under the prevailing circumstances, even if the latter completely disappear temporarily, see Strasburger (1876b:37).

28. For the sake of brevity, I use the term "dissolve," though it is possible that there is a simple loosening and distribution of the membrane particles. Incidentally, I have spoken of this previously.

29. After the ability to grow was abolished, the plasma membrane was too easily burst when it was attempted to contract it, and thus I was unable to observe how the wrinkling would develop in this case. Obviously, an artificial precipitation membrane will form wrinkles if prcesses analogous to those considered here are not involved.

30. Growth in surface area at the same speed by intussusception (not by eruption) is possible with inorganic cells of tannic gelatin glue, by the way.

31. Such proteins might possibly be in solution in the living protoplasm, and might be precipitated only when it is killed. Then the manner in which living cells are treated might affect the solubility of the proteins. See also deVries (1871b:18,31 of the reprint).

32. Cell wall, too, would not be dissolved by the above-named reagents, while on the other hand it is dissolved by and in plant organisms. This occurs, for example, before fungi penetrate (an organism), in certain germination processes, and during the transition of the sclerotia of myxoycetes into mobile plasmodia.

33. See Aronstein (1873:82) and Heynsius (1874:528). If one sucks a small amount of egg white into a glass capillary, and then dips the capillary tip in water, a membrane appears to form around the emerging drop; actually it is not a closed membrane, but a loose aggregate of fragments of the membranes that interlace the egg white of bird eggs. Thus, when drops of egg white are immersed in water, aggregates of fine membranes are formed; Monoyer (1866) erroneously seems to regard then as a closed membrane that surrounds the egg white. Egg white that has the membrane removed no longer exhibits this phenomenon. A drop simply becomes cloudy when it enters the water and is disomotically distributed in it because, for the reason given in the text, only a small quantity of protein is precipitated. The removal of the membrane from the egg white can be effected faster by shaking it vigorously together with small glass splinters rather than by cutting it repeatedly with scissors (Kühne, 1868:352).

34. On the changes of chloroplasts, see Nägeli and Schwendener (1867:553).

35. Pfeffer (1873:12) - According to Hildebrandt (1863:61) the cell sap of the flowers of Strelitzia Regina and Tillandsia amoena contains colored bodies whose pigment is soluble in water. A decision would first have to be made whether the color is retained by a covering membrane or in some other way.

36. See Nägeli (1862), also Nägeli and Schwendener (1867:420). After this manuscript had been completed, the 2nd edition (1867) of Nägeli and Schwendener's book was published; in it the finer constituents of organized substance are called micelles (p. 424). Accordingly, this term designates only a special type of molecular compound (a tagma), and chemists will hardly wish to introduce a word that is reminiscent of "cell" in an extended sense for molecular compound.

37. In the opinion of Pflüger (1875:307,342,344), the actual living part of the protoplasm body should supposedly be regarded, like the active parts of a nerve, as a single giant molecule of so-called living protein or as a structure formed out of a netlike chain of such giant molecules. If we speak simply of molecular structure, this view basically differs from one that assumes that the structure is tagmatic only in one respect--the controversy whether there was linking (concatenation) of atoms or of molecules during chemical binding has still not been settled in specific cases.

38. On the netlike structure in some protoplasm bodies, see Strasburger (1876b:20).

39. See Hofmeister (1867:24, 369) and Strasburger (1876a: 287). Obviously, it was not necessary for me to consider specifically such structural relations in the hyaloplasm (or the plasma membrane) or specific formations, such as the cilia on zoospores, while I was pursuing the formation of the plasma membrane in general. Strasburger (1876b) describes a number of factual observations.

40. Sachs (1874:41) regards the surface layer as the granule-free ground substance of the protoplasm, see Strasburger (1876b:24).

41. I again stress that it is immaterial for osmotic processes and effects whether the protoplasm is covered by a real membrane or by a peripheral layer with a different cohesion.

42. See Pfeffer (1876a:125). Incidentally, sucrose also diosmoses in only very minute amounts through copper ferrocyanide membrane. See above, p. 47.

43. When simultaneously treated with a small amount of ammonia and aniline blue or litmus, these dyes penetrated the vacuoles formed from the protoplasm of Vaucheria or Hydrocharis as little as they do normally.

44. See p. 45. Valve-like devices could also help drive liquid in one direction, but not in the opposite direction, through a membrane by means of pressure. Meckel, quoted in Ranke (1872:122), claims that something of this nature is true for the pores in the egg shell's membrane, where liquid easily filters from the shell side into the egg white side, but not the other way.

45. I take this word in its widest meaning, including precipitation through withdrawal of the solution medium, etc.

46. Obviously, such an accumulation of any substance inside the wall or in any other place is equally possible.

47. See Emmerling (1872) and Emmerling's (1874) habilitation thesis--Emmerling determined that nitric acid had been partially driven out, by a method based on diffusion, while other researchers used shaking-out methods for the same purpose. By such methods, which cause the solutes to separate, it is obviously impossible to decide how much acid is driven out when the solutions are simply mixed. To establish the simple fact that acid has been driven out, diosmotic methods could certainly be used successfully in certain cases.

48. Perhaps the traces of solution I observed on calcium oxalate crystals are produced by small amounts of inorganic acids. It must still be decided whether calcium oxalate is also dissolved in the plant in a larger quantity, see Pfeffer (1872:528). Free hydrochloric acid, incidentally, occurs in the gastric juice of animals and, as far as I know, not enough has been learned about what produces it.

49. The comparatively rapid diosmosis of many acids and alkalis is noteworthy, and helps explain why cell sap is colored red or blue, respectively.

50. A. Mayer (1875:438) recently stated that green plants can split off oxygen from other material as well as from carbon dioxide. Of course, this sort of thing is not impossible, but there is a more obvious interpretation to account for the data on which Mayer bases his assertion, an interpretation which is far more probable in this case. For the oxygen could also come from carbon dioxide that was not present as such in the tissue, but was formed by the splitting of some substance in sunlight and processed, immediately after forming, in the chlorophyll body [*Chlorophyllapparat*]. The light dependent oxidation of oxalic acid to carbon dioxide, which I mentioned in this text, shows that the formation of oxygen observed by Mayer is compatible with a simultaneous deacidification, see Becquerel (1868:60) and A. Müller (1873:25). I mention the latter because it also gives a bibliography on the effect of light on other acids. I simply wanted to show here that Mayer's actual observations do not justify the conclusions he reached, and can absolutely not do so until it has been proved beyond any doubt that no carbon dioxide is produced by sunlight in the experimental material. Such a negative reaction cannot (as I do not need to explain expressly) be alleged on the basis of observations carried out with other plants, and there is also no need for us to cite the fact that oxalis leaves do not form oxygen in the light, although they are rich in oxalic acid. From a physiological point of view, it is of secondary importance from what material the carbon dioxide was formed and under what conditions the splitting of the material in question by sunlight is possible. Also, we must not forget that carbon dioxide is actually formed by the splitting of organic

substances during inner respiration and that it could be increased, or even initiated, if the intramolecular movement of a substance was increased by light. Incidentally, Schultz and Flourens (1844) already stated some time ago that plants can produce oxygen from various organic acids; of course, they based this statement on very inadequate experiments, as Boussingault (1844) was easily able to prove.

Following this discussion of the significance of dissociation processes, I would like to say a few words on the production of organic substance in the chlorophyll body. Careful considerations that I do not wish to discuss here make it likely, to me, that the development of oxygen during assimilation is a result of a dissociation process caused by light, and that the resulting reduced substance, as it oxidizes, causes organic substance to form from carbon dioxide and water. It seems likely to me that chlorophyll itself, or at least a substance related to it, is the substance being dissociated, which immediately attracts oxygen again from carbon dioxide and water. As this undoubtedly very complex process continues, chlorophyll plays a role in the production of organic substance analagous to that played by sulfuric acid in the production of ether from alcohol. This also explains why chlorophyll is destroyed in intensive light; and when this dissociation is not total, there is an analogy for this in the fact that many dissociation processes only proceed to a certain degree.

Sachsse (1876:550) recently again expressed the view that chlorophyll itself is a product of assimilation, and that carbohydrates are produced from the substances of chlorophyll by further change and reduction. This view is incompatible with certain physiological facts. Incidentally, this view in no way explains how light is transformed into chemical power. Here, precisely, lies an important factor for the assimilation process, which can eventually be explained, even though we now have insufficient chemical knowledge about the substances involved. It must be remembered that, as we see it, only one process--dissociation--is necessarily caused by light. Such dissociation, with splitting off of oxygen, is also produced when, e.g., light is partially able to decompose mercuric oxide by splitting off oxygen (Becquerel, 1868:69). With the help of the thus reduced substance (Hg_2O?), organic compounds could probably be produced from inorganic carbon compounds and water, as well as with the help of other oxidizable substances. At least that is how we visualize the principle of the assimilation process, by means of the example of known chemical processes.

51. Nothing is known about the influence of hydrostatic pressure on the solubility of solids. Because of the complex molecular actions involved here, we probably cannot predict the results to be expected. On the other hand, we can evaluate the absorption of a gas according to Henry-Dalton's law if the gas is produced in the cell. Only in this case can a greater accumulation of dissolved gas be expected, and the diosmotic exchange with the outside liquid, which adjoins the protoplasm body and contains less gas, will continuously act to decrease the quantity of dissolved gas contained in the cell and to establish the usual osmotic state of equilibrium. When a solution is dilute, the percentage of the substance that is in a dissociated state also decreases. Naumann (1876-77:547) has compiled all the information known at this time.

52. (Hofmeister, 1867:18) --Didymium serpula, at ten mm/min, produced the highest velocity; the movements in the filament hairs of Tradescantia do not even reach one mm/min.

53. Specific studies would probably be able to give more information on the state of cohesion of the living protoplasm. For example, I would like to remind the reader that a certain cohesion is inevitable in primordial cells if the resistance of the peripheral layers is shown to be infinitesimal and, at the same time, it is proved that there is a limited osmotic pressure.

54. Such a netlike structure has been observed in isolated cases, see Strasburger (1876a:20 and 1876b:20).

55. Velten (1876) assumes that protoplasm contains a substance in a solid aggregate state. The author did not give the reasons for this assumption in the only preliminary paper I have seen so far.

56. In some botanical papers, the assumption that colloids produce high turgor is dragged along like a dogma. This has never been proved, however.

57. DeVries (1871a:7 of the reprint) calculated the concentration of the solution, for a series of salts in red beet cells, that would be required to produce the same contraction. I do not have enough data to judge whether these solutions, which produce the same effect in the plant cell, would cause an equal pressure in copper ferrocyanide membrane, for instance. It seems to me (though my reasoning is not flawless) that in the plasma membrane the difference in concentration of potassium nitrate and sucrose solutions, which have the same osmotic effect, is less than in copper ferrocyanide membranes.

58. The osmotic pressure in filament cells of Cynara scolymus is assuredly higher than one atmosphere, and I reduced the far higher pressure found by direct measurements to the above value on the basis of considerations having to do with the volume increase of cells expanded by force (Pfeffer, 1873:124). It is true these considerations were valid, and a decrease of turgor with expansion is necessary if osmotic pressure, due to water uptake, does not increase again during and after the expansion. This will probably be the case, since the cell walls, imbibed with water, can deliver it, and uptake of water occurs rapidly. That is why, during the expansion of the filaments, turgor probably does not decrease as much as I formerly assumed.

59. The resistivity to pressure in hollow vessels made from solids, for a given wall thickness, will be able to reach a maximum, as Mendelejew(1874) also discovered for glass tubes. But there is no such maximum for walls in which the tension caused by bending is almost completely balanced out in some way--by the specific molecular structure or by growth.

60. Under the above conditions, if a more dilute solution of a substance was sufficient for the same degree of contraction, this would indicate that the pressure had dropped.

61. Electric, capillary and other forces may, of course, trigger osmotic effects inasmuch as they cause chemical processes.

62. See Note 58, Part II.

63. This error was based on an incorrect interpretation of how maximal osmotic effect is produced. Even in a preliminary paper, I still subscribed to that erroneous view, and thus incorrectly interpreted the facts known about the mechanism of stimulation. See Note 57, Part I.

64. Here I mean simple diosmotic exchange where no chemical process comes about as a result of this exchange.

65. The mechanical cause for volume fluctuation in pulsating vacuoles may likely also be sought in altered osmotic effects.

66. The cells that are sensitive to stimuli are contractile organs, the reason for whose contraction I just explained. The term contractility implies only that there are dimensional changes in relatively resistant bodies when, in accordance with the original concept, we use the term simply to describe solid particles coming closer to one another. Obviously, by saying that a substance has contractility, we do not explain the cause for its change of dimension; these causes still have to be discovered. Nevertheless, and although this error has been pointed out more than once (Hofmeister, 1867:61), we still occasionally find works whose authors believe they have explained the mechanical cause of changes in shape if they state that an organ is contractile. Also, without any justification, muscle contraction and stimulus mechanics seem to be thought of, again and again, as analogous processes. Munk (1876:113) in his refutation of Sanderson was the last to explain why this view is erroneous.

67. Here I assume that the cause of stimulation movement

will be found in the osmotically effective solution, or in changes in the solution. Incidentally, the conclusions that follow can easily be carried over if the plasma membrane is found to be the controlling factor.

68. A few somewhat cursory experiments produced no results. I shall mention the fact that response to stimulus in Cynareae immediately disappears when oxygen is removed. Unfortunately, this negative result does not refute the existence of a substance that decomposes without using oxygen. The fact that NI_3 was exploded by high frequency sounds from a violin (Champion and Pellet, 1873) suggested an experiment where I attached Centaurea filaments to the strings of a violin, but I found they did not respond to vigorous bowing. Still, sounds that have a higher frequency might produce positive results.

69. See Pfeffer (1873:157). The remark of Nägeli and Schwendener (1867:374) that water cannot be conducted sufficiently rapidly in tissues that have no interstices is directly refuted by observations on the filaments of Berberis, where water may rapidly spurt from a cut surface. True, the cell walls here are quite thick, but it is possible to show that thinner cell walls also can conduct water as fast as required by a stimulation movement that does not occur too rapidly.

70. The form changes that occur here and to which Darwin seems to attach special importance are the same, after all, as those that can be observed under particular circumstances during the precipitation of certain amorphous bodies, when these are precipitated as a viscous mass.

71. This is shown by the fact that when these precipitated bodies are immersed in alcohol they pass into the coagulated state, and are now insoluble in dilute potassium hydroxide. Incidentally, this insoluble form must also be easily formed in water under certain circumstances, but I am not sure under what conditions. As for the rest, the precipitated bodies react like proteins with the usual reagents (see F. Darwin, 1876:315).

72. Here, I go into more detail than is necessary because

both Darwins make no definite statements about the continued existence of the protoplasm body after the definitive clumping, and because C. Darwin (1876:37) thinks that Heckel perhaps observed a similar phenomenon in Berberis, while Heckel expressly believes -- incorrectly -- that the protoplasm contracts. See Pfeffer (1875e and 1876b:10).

73. According to C. Darwin (1876:37), a limited precipitation occurs when the hairs are cut just below the glands. When I cut the glands a little further down, no precipitation generally took place in any of the cells.

74. Ammonia produces the same type of conspicuous change of form in the protoplasm body as do low temperature and other influences, without at first causing it to die. See Nägeli and Schwendener (1867:398) and Hofmeister (1867:53).

75. I was forced to reject the idea that the stimulus was transmitted in another way on the basis of special studies (Pfeffer, 1873-74).

76. It was not necessary for me to allude specifically to a case where two releasing forces act simultaneously. Also, I did not make a point of stating that transduction may influence metabolism, and vice versa.

77. On the expansion and expandability of bodies at different temperatures see, among others, the general discussion in Clausius (1876:199). Specifically, on the conspicuous reaction of india rubber, see also Pfaundler*(1874:62).

78. Bourdon's aneroid barometer and Fick's spring kymograph are based on this principle.

79. We should also consider if this is also the possible cause of the movements of Oscillaria, etc.

80. Thus the spores of some plants and the gemma buds of Marchantia never develop in the dark. We must also mention that the cotyledons of some plants do not continue to grow in the dark, although filled with nutrients.

81. Of course we know about movements of the protoplasm caused by uni-directional light, see Sachs (1874:721).

82. Under this heading fall combinations such as those of heliotropic and geotropic curving. I have not taken these into consideration here and in what follows.

83. Such an assumption underlies Wolkoff's hypothesis which, in fact, however, is not sufficient to explain negative heliotropism, see Sachs (1874:810).

84. Traube (1875:67) and others after him have tried to explain geotropism from the same simple causes that produce a certain kind of upward growth in inorganic cells. Since I feel that adequate knowledge of the known facts about geotropism must immediately show that this simple explanation is abslutely not adequate, I do not consider it necessary here to refute such views specifically. I cannot discuss in detail here a work by C. Kraus (1876), which I received only after finishing this manuscript. The explanation of the mechanical cause of geotropism, basically, is this --the increased growth of the surface area of the cell walls on the underside of an organ supposedly produces negative geotropism, while increased resistivity (e.g., by the thickening of the cell walls on the underside) is suppose to produce positive geotropism. This idea is not at all new, and it does not help science if this claim is based on the flat assertion: "It is undoubtedly true that gravity produces a concentrating of the cell sap that increases as you go from the upper to the lower part in a horizontally positioned root, the same way as in an artificial Traubean cell" (Kraus, 1876:440). It is totally erroneous that anything could be "undoubtedly true" as stated above and, as well, the concentration alone is no sure indicator for growth. Moreover, there exists negative geotropism without growth, and the powerful expansive force cannot be produced in the simple manner as the upward growth in an artificial Traubean cell. It would take too long to uncover all the basic errors that C. Kraus has been guilty of in interpreting the facts and in his misinterpretation of the physical aspect of certain phenomena.

85. See the corresponding chapter in the textbook by Sachs (1874), and also Pfeffer (1875d:141).

86. The distribution of bodies under the influence of gravity could also be of importance here. There has been no study so far, for cells covered by a cell wall, whether gravity induces specific movements in the protoplasm. Of course, according to Rosanoff (Sachs, 1874:813), the plasmodia of Aethalium are said to be caused to move upward by gravity, but this must at least be carefully rechecked, since what I have observed in Aethalium allows a different interpretation than the influence of gravity.

87. Precipitation membranes are pressed through the meshes of very closely woven fabrics, even at very low pressure, and emerge on the other side as sacs (see p. 2). This process is very reminiscent of the way nets are formed.

88. Reinke (1875:435) claims that the basic condition for the formation and growth of a precipitation membrane is the simultaneous contact of two membrane-forming solutes, or the existence of these on both sides of a membrane. However, in his first work, Traube (1867) showed that even bodies in pure water can become covered with a precipitation membrane (see above, p. 146). Incidentally, the formation of the cell wall as it relates to the protoplasm body needs to be investigated again.

89. If, without further ado, the similarity in structure of inorganic and living cells is invoked in order to try and explain the specific growth processes in the organism, this essentially recalls the alchemists centuries ago marvelling at the so-called metallic trees, which were also, in part, inorganic cells, see Note 5, Part I and also see Wiegleb (1790:130). History will judge our efforts as we judge analogous efforts by the alchemists. Obviously, it is quite aceptable to deduce individual determining factors from the behavior of inorganic cells, as Traube did.

90. Regarding Rosanoff's statements, see Note 86, Part II.

91. It would amount to the same thing if unequally osmotically effective solutions of different substances were present.

92. Such a constant water flow would eventually be possible even if the upper membrane, through which the flow passes from the inside to the outside, had the greatest filtration resistance. But even then, a perpetuum mobile would not have been invented (see Mayer, 1871:332). The work of N.J.C. Müller (1876:268) shows inexplicable confusion regarding osmotic effects in the organism. Thus his statement that the osmotic water flow must be proportional to the osmotically effective substance formed in the leaves is incorrect, if only because the osmotic effect depends on the specific properties of substances formed by metabolic processes and is obviously not determined by mass and by its potential energy, measured in thermal units. But the above assertion contains an even more serious error, which the following demonstrates more clearly. On p. 226, Müller states, "The real finite increase in osmotic tension in the chain is totally independent of the root, which is why an increase in the force that is maintaining the flow will never come from the root." Apart from the fact that the root, too, does absorb substances from the soil, the energy used to transport a substance and the osmotic effect after the substance has been suitable transformed may, obviously, have no relationship. For example, in the migration of starch, the material produced in leaves will be able to reach the root without any noticeable osmotic tension having formed on the way there. But the osmotic tension may reach enormous values if a sufficient quantity of sucrose forms from starch inside a cell. Obviously, this also allows so-called root force to be produced if conditions for the uni-directional secretion of water exist in the cells involved. After all, it takes comparatively little energy to take gun powder from a factory into a neighboring building, ignite it with a spark and have this triggering action produce an enormous amount of energy. Something similar occurs in an osmotic process, e.g., one that is initiated by the chemical transformation of a substance. In other types of processes, too, a triggering process can determine where work will be done. The specific formation and localization of the resulting effects in the organism is based precisely on the interaction of transduction and metabolism. When Müller overlooked this interaction, he was guilty of a fundamental error, not only in evaluating osmotic processes (as was shown before), but

in other chapters as well, where the general principles regarding relations betwen the place of origin of the chemical potential energy and the distribution of the work in the organism are developed from physical principles without regard for physiology.

93. This could be a glass tube divided by membranes into a number of chambers.

94. Hales observed extruded sap rising up to 36 feet; according to Clark (1875:559), a birch supposedly produced a pressure of 77 feet. In grapevine, Unger (1857:444) found the specific gravity of sap during bleeding to be between 1.0001 and 1.0012. The specific gravity here was found to be lower for sap that was obtained from higher up on the stem.

95. A dahlia that had been topped continued bleeding very profusely, as could have been predicted, after it was turned upside down, but I made no comparative measurements of the amounts of water outflow in this position and in the vertically upright position.

References

Aronstein, B. (1873) Pflüger's Archiv, 8, 75-93.

Askenasy, E. (1874) Bot. Zeitung, 32, --.

Baranetzky, J. (1872) Pogg. Ann. d. Physik, 147, 195-245.

deBary, A. (1864) Die Mycetozoen, 2nd ed., Engelmann, Leipzig.

Becquerel, E. (1868) La lumiere ses et ses effets, vol. II, Didot, Paris.

Beilstein, F. (1856) Ann. Chem. u. Pharm., 99, 165-197.

Berthelot, M. and L. St. Martin (1872) Ann. chim. et phys., 26, 433-462.

Berthollet, C. (1803) Essai de statique chimique, Didot, Paris.

duBois-Reymond, E. (1858) Pogg. Ann. d. Physik, 104, 193-234.

duBois-Reymond, E. (1870) Pogg. Ann. d. Physik, 139, 262-275.

Boussingault, J. (1844) Compt rendu, 19, 945-948.

Brücke, E. (1843) Pogg. Ann. d. Physik, 58, 77-94.

Brücke, E. (1861) Akad. Wiss. Wien Sitzungsb., 44, 381-406.

Budde, E. (1872) Pogg. Ann. d. Physik, 144, 213-219.

Budde, E. (1873) Pogg. Ann. d. Physik, Ergänzungsband VI, 477-498.

Buff, H. (1871-73) Lehrbuch d. physikalischen Mechanik, part 2, Vieweg, Brunswick.

Champion, P. and H. Pellet (1873) Pogg. Ann. d. Physik, Ergänzungsband VI, 174-175.

Cintolesi, F. (1876) see review in Der Naturforscher, 9, 299-300.

Clark, W. (1875) Flora, 33, 555-560.

Clausius, R. (1867) Abhandlung über d. Mechanische Wärmetheorie, 1st ed., vol. II, Vieweg, Brunswick.

Clausius, R. (1876) Abhandlung über d. Mechanische Wärmetheorie, 2nd ed., vol. I, Vieweg, Brunswick.

Clerk-Maxwell, J. (1876), see review in Der Naturforscher, 9, 422-423.

Cloetta, A. (1851) Dissertation, Diffusionsversuche durch Membranen mit zwei Salzen, Kiesling, Zürich.

Darwin, C. (1876) Ger. transl. by J. Carus, Insectenfressende Pflanzen, Schweizerbart, Stuttgart.

Darwin, F. (1876) Qtr. J. Micr. Sci., 16, 309-319.

Dutrochet, H. (1837) Memoires pour servir ..., vol. I, Bailliere, Paris, (page citations are to the Brussels edition).

Eckhard, C. (1866) Pogg. Ann. d. Physik, 128, 61-100.

Emmerling, A. (1872) Ber. Deutsch. chem. Gesellschaft, 5, 780-785.

Emmerling, A. (1874) Habilitation thesis, Kiel.

Exner, F. (1874) Akad. wiss. Wien Sitzungsb., 70, 465-501.

Feddersen, W. (1873) Pogg. Ann. d. Physik, 148, 302-311.

Fick, A. (1855) Pogg. Ann. d. Physik, 94, 59-86.

Fick, A. (1857) in J. Moleschott's Untersuchungen zur Naturlehre d. Menschen u. Thiere, vol. III, Meidinger, Frankfurt.

Fick, A. (1866) Medicinische Physik, 2nd ed., Vieweg, Brunswick.

Graham, T. (1851) Ann. Chem. u. Pharm., 77, 56-89.

Graham, T. (1854) Phil. Trans., 144, 177-228.

Graham, T. (1862) Ann. Chem. u. Pharm., 121, 1-77.

Graham, T. and F. Otto (1863) Lehrbuch d. Chemie, 4th ed., vol. II--anorgan. Chemie, Vieweg, Brunswick.

Hanstein, J. (1870) Sitzungsb. niederrhein. Gesellschaft f. Natur u. Heilkunde in Bonn--in Dechen, Verhandl. Naturhist. Vereines d. preuss. Rheinlande u. Westfalens, 27, 217-233.

Hanstein, J. (1872) Sitzungsb. niederrhein. Gesellschaft f. Natur u. Heilkunde in Bonn--in Dechen, Verhandl. Naturhist. Vereines d. preuss. Rheinlande u. Westfalens, 29, 163-166.

Hartig, T. (1843) Beiträge zur Entwickelungsgeschichte d. Pflanzen, Förstner, Berlin.

Hartig, T. (1844) Das Leben d. Pflanzenzelle, Förstner, Berlin.

Hartig, T. (1858) Entwickelungsgeschichte d. Pflanzenkeims, Förstner, Leipzig.

Heynsius, A. (1874) Pflüger's Archiv, 9, 514-552.

Hildebrandt, F. (1863) Jahrb. f. wiss. Bot., 3, 59-76.

Hirzel,* (1868) Dingler's Polytech. Journ., 141, --.

Hofmeister, W. (1858) Flora, 26, 1-12.

Hofmeister, W. (1867) Die Lehre von d. Pflanzenzelle, Engelmann, Leipzig.

Hoppe-Seyler, F. (1866) Medicinisch-chemische Untersuchungen, vol. I, Hirschwald, Berlin.

Jagielski, (1859) Program d. Gymnasiums zu Trzemeszno.

Jolly, P. (1849) Ztschr. f. rat. Med., 7, 83-148.

Kekülé, A. (1861) Lehrbuch d. organ. Chemie, Enke, Erlangen.

Kohlrausch, F. (1857) Gesellschaft Beförderung gesammten Naturwiss. Marburg, 8, 1-88.

Kopp, H. (1847) Geschichte d. Chemie, vol. IV, Vieweg, Brunswick.

Kraus, C. (1876) Flora, 34, 438-445.

Kraus, G. (1869-70) Jahrb. f. wiss. Bot., 7, 209-260.

Kühne, W. (1864) Untersuchungen über d. Protoplasma, Engelmann, Leipzig.

Kühne, W. (1868) Lehrbuch d. Physiologischen Chemie, Engelmann, Leipzig.

Kürschner, (1842) in Wagner's Handwörterbuch d. Physiologie, vol. I, 35-75, Vieweg, Brunswick.

Liebig, J. (1848) Untersuchungen über einige Ursachen ..., Vieweg, Brunswick.

Link, H. (1833) Pogg. Ann. d. Physik, 27, 193-239.

Loschmidt, J. (1870) Akad. wiss. Wien. Sitzungsb., 61, 367-380.

Ludwig, C. (1849) Ztschr. f. rat. Med., 8, 1-52.

Marignac, C. (1874) Ann. chim. et phys., ser. V, 2, 546-581.

Matteucci, C. and A. Cima (1845) Ann. chim. et phys., 13, 63-86.

Mayer, A. (1871) Lehrbuch d. Agriculturchemie, vol. I, Winter, Heidelberg.

Mayer, A. (1875) Landwirth. Versuchsstat., 18, 410-452.

Mendelejew, D. (1874) Ber. Deutsch. chem. Gesellschaft, 7, 126-127.

Meyer, L. (1872) Die modernen Theorien d. Chemie, 2nd ed., Maruschke and Berendt, Breslau.

Meyer, O. (1874) Pogg. Ann. d. Physik, 153, 619-621.

Mohl, H. (1844a) Bot. Zeitung, 2, 289-294.

Mohl, H. (1844b) Bot. Zeitung, 2, 273-277.

Mohl, H. (1844c) Bot. Zeitung, 2, 305-310.

Mohl, H. (1846) Bot. Zeitung, 4, 73-78.

Mohl, H. (1855) Bot. Zeitung, 13, 689-701.

Monoyer, F. (1866) Soc. chim. France Bull., Paris, 5, 444.

Mousson, A. (1871) Pogg. Ann. d. Physik, 142, 405-417.

Mousson, A. (1871-74) Die Physik, 2nd ed., vol. I, Schulthess, Zürich.

Müller, A. (1873) Ueber d. Einwirkung d. Lichtes auf Wasser, Schmidt, Zürich.

Müller, H. (1876) Flora, 34, --.

Müller, N. (1876) Botanische Untersuchungen, Vol. I, Winter, Heidelberg.

Munk, H. (1876) Die elektrischen u. Bewegungserscheinungen ..., Kurtz, Leipzig.

Nägeli, C. (1844) Ztschr. f. wiss. Bot., 1, 1-188.

Nägeli, C. (1855) in Pflanzenphysiologische Untersuchungen, part 1, ed. C. Nägeli and C. Cramer, Schulthess, Zürich.

Nägeli, C. (1858) in Pflanzenphysiologische Untersuchungen, part 2, ed. C. Nägeli and C. Cramer, Schulthess, Zürich.

Nägeli, C. (1862) Sitzungsb. Akad. Wiss. München, part I, 290-324, also reprinted in Nägeli, C. (1863) Botanische Mittheilungen, vol. I, p. 203, Straub, Munich.

Nägeli, C. and S. Schwendener (1867) Das Mikroskop, Engelmann, Leipzig.

Naumann, A. (1869) Grundriss d. Thermochemie, Vieweg, Brunswick.

Naumann, A. (1872) Über Moleculverbindungen nach festen Verhältnissen, Winter, Heidelberg.

Naumann, A. (1873-75) in Gmelin-Kraut's Handbuch d. Chemie, 6th ed., vol. I, part 1, Allgemeine u. physikal. Chemie, p. 1-320, Winter, Heidelberg.

Naumann, A. (1876-77) in Gmelin-Kraut's Handbuch d. Chemie, 6th ed., vol. I, part 1, Allgemeine u. physikal. Chemie, p. 321-887, Winter, Heidelberg.

Neumann, C. (1872) Ber. Kgl. Sächs. Gesellschaft Wiss., 24, 49-64.

Pfaundler, L. (1874) Pogg. Ann. d. Physik, Jubelband, 182-198.

Pfaundler,* L. (1874) Pogg. Ann. d. Physik, 153, --.

Pfeffer, W. (1872) Jahrb. f. wiss. Bot., 8, 429-574.

Pfeffer, W. (1873) Physiologische Untersuchungen, Engelmann, Leipzig.

Pfeffer, W. (1873-74) Jahrb. f. wiss. Bot., 9, 308-326.

Pfeffer, W. (1875a) Bot. Zeitung, 33, 734.

Pfeffer, W. (1875b) Situngsb. niederrhein. Gesellschaft f. Natur u. Heilkunde in Bonn, Aug. 2, 1875-in Dechen, Verhandl. Naturhist. Vereines d. preuss. Rheinlande u. Westfalens., 32, 276-279.

Pfeffer, W. (1875c) Situngsb. niederrhein. Gesellschaft f. Natur u. Heilkunde in Bonn, July 5, 1875-in Dechen, Verhandl. Naturhist. Vereines d. preuss. Rheinlande u. Westfalens., 32, 198.

Pfeffer, W. (1875d) Die Periodische Bewegungen d. Blattorgane, Engelmann, Leipzig.

Pfeffer, W. (1875e) Bot. Zeitung, 33, 289-291.

Pfeffer, W. (1876a) Landwirtsch. Jahrb., 5, 87-130.

Pfeffer, W. (1876b) Bot. Zeitung, 34, 9-13.

Pflüger, E. (1875) Pflüger's Archiv, 10, 251-367.

Poiseuille, J. (1843) Ann. chim. et phys., ser. III, 7, 50-74.

Pringsheim, N. (1854) Untersuchungen über d. Bau u. Bildung d. Pflanzenzelle, Hirschwald, Berlin.

Quincke, G. (1869) Pogg. Ann. d. Physik, 137, 402-414.

Quincke, G. (1870) Pogg. Ann. d. Physik, 139, 1-89.

Rammelsberg, C. (1863) in Graham-Otto's Lehrbuch d. Chemie, 4th ed., vol. II, anorgan. Chemie, Vieweg, Brunswick.

Ranke, J. (1872) Grundzüge d. Physiologie d. Menschen, 2nd ed., Engelmann, Leipzig.

Regnault, V. (1847) Mem. acad. sci. inst. France, Paris, 21, 1-767.

Reinke, J. (1875) Bot. Zeitung, 33, 425-437.

Sachs, J. (1865) Handbuch d. Experimental physiologie d. Pflanzen, Engelmann, Leipzig.

Sachs, J. (1872) Arbeiten Bot. Inst. Würzburger, 2, --.

Sachs, J. (1873) Lehrbuch d. Botanik, 3rd ed., Engelmann, Leipzig.

Sachs, J. (1874) Lehrbuch d. Botanik, 4th ed., Engelmann, Leipzig.

Sachsse, R. (1874) Chem. Centralblatt, 5, 237-239.

Sachsse, R. (1876) Chem. Centralblatt, 7, 550-552.

Schmidt, W. (1856) Pogg. Ann. d. Physik, 99, 337-388.

Schmidt, W. (1857) Pogg. Ann. d. Physik, 102, 122-167.

Schultz, and P. Flourens (1844) Compt. rendu, 19, 524-525.

Schumacher, W. (1861) Die Diffusion in ihren Beziehungen zur Pflanze, Winter, Heidelberg.

Strasburger, E. (1875) Zellbildung u. Zelltheilung, Dabis, Jena.

Strasburger, E. (1876a) Zellbildung u. Zelltheilung, 2nd ed., Dabis, Jena.

Strasburger, E. (1876b) Studien über d. Protoplasma, Dufft, Jena.

Thomson, W. (1871) Ann. Chem. u. Pharm., 157, 54-66.

Traube, M. (1866) Centralblatt f. Med. Wiss., 4, 97-100.

Traube, M. (1867) Archiv Anat. u. Physiol., 87-165.

Traube, M. (1874) Gesellschaft Deutscher Naturforscher u. Aertze, Tageblatt d. 47. Versammlung in Breslau, 191-199.

Traube, M. (1875) Bot. Zeitung, 33, 56-70.

Unger, F. (1857) Sitzungsb. Kaiserl. Akad. Wiss. Wien. Math.-Naturw. Cl., 25, 441-470.

Velten, W. (1876) Bot. Zeitung, 34, 328-332.

Vierordt, K. (1846) Archiv f. physiol. Heilkunde, 5, 479-517.

Vierordt, K. (1848) Pogg. Ann. d. Physik, 73, 519-531.

Voit, E. (1867) Pogg. Ann. d. Physik, 130, 227-240 and 393-423.

deVries, H. (1871a) Arch. Neerl., 6, 117-125.

deVries, H. (1871b) Arch. Neerl., 6, 245-286.

Wiegleb, J. (1790) Geschichte d. Wachsthums ..., vol. I, Nicolai, Berlin.

Wilhelmy, L. (1864) Pogg. Ann. d. Physik, 122, 1-17.

Wüllner, A. (1870) Lehrbuch d. Experimentalphysik, 2nd ed., Teubner, Leipzig.

Zincke, T. (1876) Ann. d. Chem., 182, 241-246.

Index